DIGITAL DC

DIGITAL DOMESTICITY

Media, Materiality, and Home Life

Jenny Kennedy, Michael Arnold,
Martin Gibbs, Bjorn Nansen,
AND
Rowan Wilken

OXFORD
UNIVERSITY PRESS

Oxford University Press is a department of the University of Oxford. It furthers
the University's objective of excellence in research, scholarship, and education
by publishing worldwide. Oxford is a registered trade mark of Oxford University
Press in the UK and certain other countries.

Published in the United States of America by Oxford University Press
198 Madison Avenue, New York, NY 10016, United States of America.

© Oxford University Press 2020

All rights reserved. No part of this publication may be reproduced, stored in
a retrieval system, or transmitted, in any form or by any means, without the
prior permission in writing of Oxford University Press, or as expressly permitted
by law, by license, or under terms agreed with the appropriate reproduction
rights organization. Inquiries concerning reproduction outside the scope of the
above should be sent to the Rights Department, Oxford University Press, at the
address above.

You must not circulate this work in any other form
and you must impose this same condition on any acquirer.

Library of Congress Cataloging-in-Publication Data
Names: Kennedy, Jenny (Postdoctoral researcher), author.
Title: Digital domesticity : media, materiality, and home life /
Jenny Kennedy, Michael Arnold, Martin Gibbs, Bjorn Nansen, and Rowan Wilken.
Description: New York, NY : Oxford University Press, [2020] |
Includes bibliographical references and index.
Identifiers: LCCN 2019053619 (print) | LCCN 2019053620 (ebook) |
ISBN 9780190905781 (hardback) | ISBN 9780190905798 (paperback) |
ISBN 9780190905811 (epub)
Subjects: LCSH: Home automation—Social aspects. |
Technological innovations—Social aspects. |
Home economics—Technological innovations. |
Human–computer interaction—Social aspects.
Classification: LCC TK7881.25 .K46 2020 (print) |
LCC TK7881.25 (ebook) | DDC 643/.6—dc23
LC record available at https://lccn.loc.gov/2019053619
LC ebook record available at https://lccn.loc.gov/2019053620

1 3 5 7 9 8 6 4 2

Paperback printed by Marquis, Canada
Hardback printed by Bridgeport National Bindery, Inc., United States of America

CONTENTS

List of Figures   vii
Acknowledgments   ix
Project Legend   xi

Introduction   1

1. Histories   16

2. Ecologies   56

3. Appropriations   86

4. Housekeepings   127

5. Negotiations   164

6. Non-uses   201

CONTENTS

7. Displacements   237

Conclusion   269

References   *281*
Index   *301*

FIGURES

1.1.	Key4IT Help Page interface	27
1.2.	Installing fiber-optic cable	40
1.3.	NBN household connection hardware	47
2.1.	Probe pack for Connected Homes project	75
2.2.	Connected Homes junk-mail catalog	78
2.3.	EthnoCorder app interface	79
2.4.	Paparazzi shot, "couch office"	80
3.1.	Jeremy and Amy's study	105
3.2.	Floor plans showing removal of media room	107
3.3.	Toddler-proof fence excluding the TV from the living area	113
3.4.	Installed NBN household connection hardware	122
4.1.	John's home entertainment system configuration	135
4.2.	Example of labor gone into the material management of household	148
5.1.	Children engaging with individual games and streaming content	178
5.2.	Technology creep	192

FIGURES

6.1.	Rooftop wireless broadband internet connection in rural home	213
6.2.	Scanner used for document storage	233
7.1.	Dumped CRT television	242
7.2.	Broken mobile phone	261
7.3.	Craig's garage with piles of retained technology	263

ACKNOWLEDGMENTS

Many people have contributed to the successful completion of this book, and to the research projects and outputs that have fed into this book. We would like to thank the following people, all of whom have helped at different points along the way and in a variety of capacities: Tom Apperley, Craig Bellamy, Marcus Carter, Brent Coker, Marcos Dias, Hilary Davis, John Downs, Sophie Freeman, Robbie Fordyce, Mitchell Harrop, Paul Hill, Steve Howard, Catherine Howell, Tamara Kohn, Dale Leorke, Kate Mannell, Karen Mecoles, Scott McQuire, James Meese, Sora Park, Sonja Pedell, Matthew Riddle, Melissa Rogerson, Christopher Shepherd, Larry Stillman, Julian Thomas, Emily van der Nagel, Frank Vetere, Hannah Withers, and Philippa Wright.

At Oxford University Press, we wish to thank Sarah Humphreville, Bronwyn Geyer, and Hannah Doyle. We also wish to thank Hallie Stebbins for her initial and enthusiastic support of this project.

We also wish to thank the many householders and technological innovators who, over many years, have given generously of their

ACKNOWLEDGMENTS

time, have shared their insights and experiences, and have been active research participants in the numerous projects that inform this book.

Closer to home, Jenny would like to thank Evelyn; Mike would like to thank Jen; Martin would like to thank Kitty, Patrick, and Ivy; Bjorn would like to give a big thanks to Lauren, Harvey, and Lulu for their love and patience; and Rowan would like to thank Karen, Laz, Max, and Sunday for their love, support, and encouragement.

This book brings together the findings from three research projects funded by the Australian Research Council (ARC), some of which have been previously published in journal form. The ARC projects were C00106911, "The Wired Homes Project"; DP0557781, "The Connected Home: Probing the Effects and Affects of Domesticated Information and Communication Technologies"; and DP130101519, "An Investigation of the Early Adoption and Appropriation of High-Speed Broadband in the Domestic Environment." We gratefully acknowledge the ARC's generous financial support. We are also grateful for the support provided by the Australian Communications Action Network, The Institute for a Broadband Enabled Society, and the Melbourne Networked Society Institute, and would like to thank a few key people at these organizations: Narelle Clark, Ken Clarke, Adam Lodders, and Thas Nirmalathas.

Jenny Kennedy, RMIT University
Michael Arnold, University of Melbourne
Martin Gibbs, University of Melbourne
Bjorn Nansen, University of Melbourne
Rowan Wilken, RMIT University

PROJECT LEGEND

Phase	Year range	Details of contributing research
Wired Homes	2002–2004	**Wired Homes: Communication and Information Technologies in a Residential Setting** Contributors: Associate Professor Michael Arnold, Associate Professor Gavan McCarthy, Dr. David Platt Funded through the Australian Research Council Strategic Partnerships with Industry—Research & Training Scheme (C00106911)

PROJECT LEGEND

Phase	Year range	Details of contributing research
Connected Homes	2004–2010	**The Connected Home: Probing the Effects and Affects of Domesticated Information and Communication Technologies** Contributors: Associate Professor Michael Arnold, Associate Professor Martin Gibbs Australian Research Council Discovery Project Award DP0557781 **The Adoption and Use of Residential Community Intranets (Springthorpe)** Contributor: Associate Professor Martin Gibbs Funded through The University of Melbourne, Early Career Researcher Award
High-Speed Broadband	2011–2017	**An Investigation of the Early Adoption and Appropriation of High-Speed Broadband in the Domestic Environment** Contributors: Associate Professor Michael Arnold, Associate Professor Martin Gibbs; Dr. Jenny Kennedy; Dr. Bjorn Nansen; Associate Professor Rowan Wilken Australian Research Council Discovery Project Award (DP130101519)

PROJECT LEGEND

Phase	Year range	Details of contributing research
		Framing the NBN: Consumer Attitudes
		Contributors: Associate Professor Michael Arnold, Associate Professor Martin Gibbs, Dr. Bjorn Nansen, Associate Professor Rowan Wilken
		Institute for a Broadband Enabled Society (IBES) Seed Funding Grant
		HSB and Home-Based Work
		Contributors: Associate Professor Michael Arnold, Associate Professor Martin Gibbs, Dr. Bjorn Nansen, Associate Professor Rowan Wilken
		Melbourne Research Office—Collaboration Grant
		High-Speed Broadband and Household Media Ecologies
		Contributors: Associate Professor Michael Arnold, Associate Professor Martin Gibbs, Dr. Bjorn Nansen, Associate Professor Rowan Wilken
		Australian Communications Consumer Action Network (ACCAN)

INTRODUCTION

What is *digital domesticity*? The term first appears in an article by Lynn Spigel titled "Media Homes: Then and Now," where she describes the "smart home" as "the contemporary model of digital domesticity" (Spigel 2001, 385). Domesticity is clearly concerned with ways of living, with dwellings, with families and housemates, and with how people live in their homes. Yet, domestic life has always been mediated by lots of "stuff"—furniture, appliances, fittings, decorations, utilities—and now increasingly involves proliferating media technologies, communication infrastructures, and software applications that reconfigure life in the contemporary home. In this book, we examine home life and how it is negotiated with, through, and around the materiality of digital technologies; how they are brought into the home; how they populate spaces within the home; how people negotiate use of spaces, as spaces and in time; how people make use of expertise within and beyond the home; and, how, over time, technologies fail and become dysfunctional, thus continuing the ever-changing ecologies of household media.

The digital domesticity approach to the home as a site for research extends beyond specific examples of specific technologies (say, television, or the internet) to encompass all of the interrelated media technologies in the home, envisaged as a whole as an "ecology"—a concept we unpack in greater detail in Chapter 2. Briefly, ecologies are concerned with the life cycles of digital media and communications technologies populating the home, the materiality of domestic architectures and spaces, the interplay of interpersonal relations and domestic technologies, and the extension of these beyond the domestic through the social and political economies of digital industries and services. In our approach, digital domesticity constitutes a complex, interactive, socio-technical ecology. And by attending to the entire domestic ecology, rather than a particular technology, we capture the interrelations of technologies with householders and with one another and provide a more comprehensive and materialist perspective to patterns of digital change within and beyond the home. Meanwhile, in attending to processes and dynamics of appropriation, maintenance, negotiation, disuse, and displacement, the ecology approach to household media moves beyond the domestication of particular technologies to situate them within larger social, commercial, political, and environmental considerations.

This book is therefore concerned with the home, but it is not bounded by the home. While the home provides a necessary anchor point for our empirical and theoretical work, we are well aware that the home is not self-contained but is a node in multiple commercial, cultural, and technical networks, all of which interact, and all with local implications and global reach. The home's socio-technical ecology operates in recursive relations with these much larger ecologies, none of which can be ignored if the home is to be understood. This book unearths this digital domesticity through accounts of evolving socio-technical relations as they unfold in processes of adopting and adapting to new innovations, using and maintaining as well as neglecting the complex of technologies in the home, and confronting the obsolescence of particular technologies and failure of systems of consumer technologies.

INTRODUCTION

OVERVIEW OF PROJECTS

The book draws on seventeen years of empirical research on domestic technologies. Through numerous research projects (detailed in the Project Legend at the front of the book), we have captured the longitudinal change that has taken place in the home since the turn of the twenty-first century, particularly the shift from wired, limited, and fixed connections to wireless, multiple, and mobile connectivity. We have identified the dynamics that energize the changing domestic media environment, and, with our household research participants, we have explored the implications of these technology changes for daily family practices, and the implications of this for our understanding of relationships among technology, the home, and everyday life.

Our analysis began at the turn of the twenty-first century as people moved into fifty-two newly constructed houses at Williams Bay, Melbourne, Australia. At a time of standalone desktop computers, the occasional dialup modem, and bulletin board discussion lists, new residents found that an advanced communications and information system was being installed as a standard household feature allowing households to create and maintain items such as community news (from local groups such as the football club and environmental groups), message boards, discussion lists, notices of upcoming events, links to the local council, local classified advertising, newsletters, a general calendar of local events and specific calendars of events for specific groups, and collections of documents (e.g., those generated by council activities). Innovations such as these, which claimed to enhance community spirit and residential lifestyle, were of course not unique to Williams Bay: well-known international examples of the day included the Digitale Stad initiative in Amsterdam, Blacksburg in the United States (Silver

2000; Jankowski 2002), Netville in Toronto (Hampton 2000; Hampton and Wellman 2000; Jankowski 2002), and the Japanese Multifunction Polis experiments (Inkster 1991). Examining how people used, adapted, and ultimately did not use the community intranet at Williams Bay, and later the surrounding suburban area, was the basis for what we dubbed the "Wired Homes Project." The community intranet was initially offered to the original fifty-two households of Williams Bay, but over the next eighteen months it was made available to other households in the broader suburban area of Williamstown (see Chapter 1).

This research project and those that followed are categorized according to three distinct phases of investigation (see the Project Legend), with some overlap in the times of data collection. Each phase is roughly aligned with significantly different "eras" of digital domesticities: (1) no networking or early dialup internet connectivity, home offices, and desktop computing; (2) fixed-line internet, mobile devices, Wi-Fi and fluid spaces; and (3) high-speed broadband, streaming services, mobile devices, multiple screening, and ubiquitous connection.

Phase 1, *Wired Homes*, ran from 2002 to 2004. Research was largely ethnographic in character, with graduate research students embedded with the developers of the community intranet (Carmody 2010) and embedded with the Williams Bay community serviced by the community intranet (Wright 2005). Data began with house-to-house questionaries of the fifty-two households who resided in Williams Bay and had access to the new community intranet. Researchers also observed activity on the community intranet, attended and participated in town meetings, joined and actively participated in the developer's Community Advisor Committee, and conducted interviews and many informal discussions with residents of the area. Intensive fieldwork was conducted between February 2002 and July 2004. The period was also marked by our close engagement with the system developers and the housing developers. The project mapped households' engagement and responses to the new communicative possibilities introduced by the community intranet, its associated functionalities, and the features of the domestic digital ecology such as early internet access.

Phase 2, *Connected Homes*, followed Wired Homes from 2004 to 2010 and involved recruiting and actively engaging with seventeen households as "co-researchers." Participants in this phase have included twelve urban households and five rural households, and a wide diversity of household types ranging from students in shared living arrangements through single-parent households to the nuclear family. Five households were from Wheatcliffs, a small rural town with a population of 450 that is four hours' drive from Melbourne (Stillman et al. 2010). Another six households were drawn from Springthorpe, a high-end residential development in the middle suburbs of Melbourne, which also featured community intranet initiatives similar to those at Williams Bay. Each household was visited at least twice by researchers, usually working in pairs. Data collection began with a "technology tour" (Nansen et al. 2016) of the home and was followed with an interview. All participants were left with "domestic probes" (Gaver, Dunne, and Pacenti 1999), a selection of tasks and activities participants were asked to perform that were intended to prompt awareness and reflection on the domestic digital ecologies of the home (Graham et al. 2007). Researchers returned a few weeks later to reinterview household members and to discuss issues, ideas, and insights generated by the domestic probes. Data collection began in March 2004 and continued through September 2005 for urban households and occurred in January and February 2007 for rural households.

Follow-up interviews were conducted with many of the households in September 2008.

Phase 3, *High-Speed Broadband*, followed the Connected Homes phase and primarily draws on a sample of twenty-four households with whom we engaged over a four-year period from September 2013 to December 2017. These households were stratified across urban, regional, and rural locations. Efforts were made to engage these households through technology tours, domestic probes using a smartphone app, and interviews conducted annually over the years 2013 to 2017. At the start of the project all of these households were newly connected to Australian's high-speed broadband network, the National Broadband Network (NBN). The aim of the longitudinal research design was to trace changes in the digital domesticities of these households as they adapted and adopted to the NBN. Phase 3 also incorporates data from a participant sample of six households in the southern island state of Tasmania, Australia (July 2010 and February 2011), twenty households living in the inner-city Brunswick area (March and May 2012), and interviews of four households living in suburban Melbourne (July and August 2013), as well as a national survey of attitudes and perceptions of the NBN (March 2013).

Methodologically, our empirical work has drawn on and developed a range of innovative ethnographic and digital approaches, such as household interviews, "technology tours," remote data collection via mobile applications, domestic probes and participant-generated data collection, to access and include householders as participants in the research process (see Chapter 2). Our intent throughout, and especially as our experience with the methods and area of investigation improved, was to enroll and cast householders as participant researchers rather than informants. This feature of our research design emphasizes the view that everyone is an expert in his or her own experience. This temporal, historicized, and relational account provides a necessary counterpoint to social accounts of domestication that elide the materiality of media themselves, and to technocentric and technophilic literature that attends to the new technology of the day in isolation, or as simply being present, erasing their antecedents, applauding their claimed functionality, and glossing the complex relations between technologies and the lives we lead with and through them.

HOW THIS BOOK IS STRUCTURED

Today the home is a fluid place that is bristling with technologies that were different yesterday and will be different again tomorrow. To contextualize this, Chapter 1, "Histories," commences with an overview of these changing domestic media ecologies since the turn of the present century. We present a longitudinal view that tracks shifts in domestic life through the changing technologies we use for communications, entertainment, home maintenance, and work and study, and we track changes in technologies as they are shaped by domestic life. The introduction situates the seventeen years of research presented in the book on a broader historical canvas of media throughout the twentieth and into the twenty-first centuries in the Anglophone world, and it presents a concise introduction to scholarship mapping the relationship between our domestic lives and our technologies. This chapter broadly tracks the processes of change as they have unfolded in the home and discusses the implications of socio-technical change across the whole range of domestic media technologies, as they flow through the home and ripple out through home life. In turn, this chapter situates this longer global history within more recent and local contexts that initiated our research agenda.

In this way, the historical account of Chapter 1 summarizes and presents previous scholarship on the domestication of media technologies, as well as mapping the shifting and accumulating hardware devices, software systems, and infrastructural layers of technology shaping contemporary digital domesticity. The domestic sphere is seen not only as a significant context for technology consumption, but as a significant, multifaceted site for making meaning of technologies, and for postmarket innovation in technology developments and application. For readers of the book as a whole, Chapter 1 provides a context to position the chapters that follow, and for the reader of this chapter alone, it provides a concise reference to the changing domestic media landscape and to the relevant literature.

Chapter 2, "Ecologies," details our theoretical approach to understanding materiality and media of the home. It presents a framework for considering how technologies do not simply serve social uses or express social values but have a presence and performativity and interact with one another and with householders, and how they are embedded within the infrastructures of the home and daily life. Chapter 2 explicitly builds on prior work in the field and extends it through critical analysis of media ecologies.

David Altheide's *An Ecology of Communication* (1995) draws our attention to a world of communicative acts that emanate from conjunctions of communications technologies, communications formats, and communicative activities. Particular conjunctions of technologies, formats, and actions are pervasive in any given situation, and the media constituents that pertain are multiple, varied, yet interdependent, hence Altheide's apt appropriation of the metaphor "ecology." The notion that domestic media technologies are usefully understood as forming an "environment" (rather than being objects that occupy an environment) is a profound one, and the fact that the claim looks somewhat unremarkable many years after publication is a tribute to its usefulness rather than a criticism. In this spirit, we situate this book's analysis in the context of Altheide's ecology of communication but also use this position—and the domestication approach to household media research—as a point of departure that intersects with the wider material turn within the humanities. We depart from Altheide's construction, and make a contribution to the development of theory, by differing from his insistence that the ecology is populated by symbolic entities, not by things. We describe our novel fieldwork approaches used to conceptually explore the "thingness" of the things that mediate the physical and spatial aspects of environments of communications technologies in the ecology of the home.

Chapter 3, "Appropriations," considers how media landscapes in the home have shifted over this century and examines how devices relate to each other and relate to householders to create dynamic and evolving media ecologies. At the turn of the twenty-first century, a typical domestic media ecology comprised a cathode-ray television in the living room displaying a limited number of free-to-air channels, and perhaps connected to a videocassette recorder; a desktop computer in a home office, perhaps connected to a dialup modem; and a landline telephone, in most cases located in a communal area in the home. In more recent years, the home has become a place for high-definition "smart" televisions, intelligent multifunction settop boxes, game consoles, digital radio, high-speed broadband, cable and wireless home networks, mobile computing, cloud connections, online government service provision, networked commerce, gesture-controlled games, extensive social networking, personal and portable entertainment systems, digital cameras, streaming media of all kinds, and home office systems and home education systems. How and why have these technologies been appropriated? How has this ongoing appropriation reconfigured the domestic media ecology and the life that is lived within this ecology?

Technologies that are appropriated are not just located in the home; in important ways they constitute a place as a home. They form assemblages that constitute the sociotechnical environment in which we live, and their particular constitution has serious implications for what goes on at home—for our day-to-day lived experience. This chapter examines the motivations and strategies for bringing new technologies into the home, exploring how new media technologies are incorporated and appropriated into everyday practices, and how everyday practices are adjusted and adapted to these new devices. It also considers the rationales for home setups, on a scale that embraces global cultural and economic forces to the particularities of locating specific devices. In doing so, this chapter both builds on and departs from prior work within media studies—such as the "domestication" approach—by moving beyond the symbolic significance of media to account for the more material presence, relations, and performance of technologies as shaping the contemporary home.

Chapter 4, "Housekeepings," focuses on the significant and often invisible forms of "articulation work" (the work to keep things working) needed to maintain digital media in good working order and fit-for-purpose in the domestic media ecology. It considers the labor of investigating options and deciding on and setting up new technologies as well as their ongoing maintenance. This chapter examines both the work and who does the work of maintaining and managing digital media. It also examines the relations of power, authority, gender, labor, and expertise that go into decision making, appropriating, maintaining, and using household digital technologies. In doing so, it furthers empirical developments concerning the notion of domestic media ecologies.

Consistent with our theoretical approach, we treat the subjects and the objects of these labors symmetrically and look not only to the work done by the householders to build and maintain a media ecology but also to the way this work folds back from the technology to build and maintain the identity of the householder.

At the turn of the century it was uncontroversial to assert that domestic technologies were gendered in their design, construction, patterns of use, and ideology, and that technologies were similarly discriminating in terms of age or generation. Today, Cynthia Cockburn and Susan Ormrod's (1993) "white goods, brown goods" gender binary and Marc Prensky's (2001) "digital natives, digital immigrants" generational binary are not nearly as crudely drawn, but it remains the case that, in any particular context, technical work done in the household and affective responses to that work are not evenly distributed; consequently, relations to technologies constitute a resource for identity construction. While differences often arise from personal preference, these preferences are often stable for long enough to help define the household's settled division of labor, and one's position in that labor market folds back to articulate a personal identity. Interestingly, this is often around markers that might apply to both the person and the technology, such as "useful," "contemporary," "reliable," and "smart" (or their antonyms). In Chapter 4, therefore, we argue that relations between individual householders and the labor associated with the appropriation, maintenance, and use of media technologies constitute important fields for the achievement and designation of identity (some would say ontology) for both the ecology and the householder.

Homes were connected electronically to the outside world less than one hundred years ago and their connection remained tenuous even at the turn of this century. And, now, as if it has not been burdened with enough responsibility, the home is asked to play a major role as a communications node in a global network of interactive media. The home now connects directly to friends, acquaintances, and a billion strangers: to the local community; to work; to social, political, and commercial organizations; and to entertainment and service providers. The domestic ecology in contemporary pictures of contemporary life is a place of leisure, a command and control center, a place for production, and a place for consumption.

In this place that is the contemporary home, parenting occurs, and, in Chapter 5, "Negotiations," we examine the various ways digital media technologies and devices are embedded and embodied in the everyday activities of parents and children working, playing, educating, socializing, and entertaining in the home. Through this century, we have visited many families and have talked with them about their experiences and their parenting strategies in the face of new technologies. In this chapter, we identify the major strategies and stances and contextualize their nuances and subtleties vis-à-vis the particulars of the family relationships. We also place our findings in the context of the literature on families and technology use, relating the particularities of the vignettes to observations derived from quantitative and larger-scale studies.

A key factor in these negotiations is an approach to time. Technologies act in time and technologies act on time and control over time, and negotiating the expenditure of time was a frequently occurring motif in our conversations with parents and children. Defining and then enacting bedtime, dinnertime, homework time, playtime, time to work, time to rest, time to sleep, and so on was far from simple in conditions where too much time is wasted here, too little time is available there, and time lurches between dragging and flying. The family home is a place that is noisy with people and noisy with incoming and outgoing media; in this chapter, we examine the variety of ways that all of this is negotiated by parents and their children in the everyday activities of working, playing, educating, socializing, and entertaining in a home ecology. Our intention here is identify the negotiated rhythms of temporal dynamics at play in the domestic media ecology, through an acoustic metaphor for the rhythms of time. Through this metaphor we identify four forms of interlaced rhythms: (1) a *polyphonic drone*, in which the media ecology is "always on, always connected" (Baron 2008); (2) a *polychromic dissonance*, meaning that routines are not organized in a linear, sequential, and ordered manner—rather, the ecology offers the potential for ordering tasks in time so that they are frequently started, stopped, switched, and interspersed with each other in a way that is frequently dissonant; (3) an *asynchronous consonance*, to impose ordered routines in the face of the drone and of dissonance; (4) and *orchestrating domesticity* through temporal technologies, such as time charts, diaries, calendars, and personal organizers of various kinds.

Domestic media ecologies are clearly shaped by the household denizens who interact with the technologies within these ecologies but are also shaped by varied modes of *not using* technologies—concerns that form the focus of Chapter 6, "Non-uses." This issue of non-use of domestic media technologies has traditionally been treated as a question of inequality and exclusion, measured in terms of a household's access to a range of resources—financial, technical, social, and human—and has been addressed as a problem of scarcity or deficit to be overcome. There is, however, a growing body of research literature attending to the not inconsiderable volume and variation of disaffected or discriminating "non-use" emerging in places of technology abundance, to which this chapter contributes.

Our research shows that intensities of non-use vary across the rhythms and spaces of family life within technologically rich houses and communities. In Chapter 6, non-uses are not a question of individual agency, nor do they occur in isolation; instead, they emerge in relation to the people, practices, and technical population (digital infrastructures, devices, services, and software) that configure the household. The modes of non-use reported on in Chapter 6 emerge most distinctly around the margins of technology design, affordance, functionality, expertise, adoption, and use.

The modes of non-use identified in Chapter 6 are characterized as "peripheral," "partial," "vicarious," or "radical." The chapter considers how these interstitial forms of non-use challenge a binary opposition to use. Forms of non-use may be driven by provisional and situated constraints or motivated by clearer social, personal, or political agendas and decisions on the part of the home dwellers. Yet, like modes of use, modes of non-use cultivate household media ecologies in ways that draw attention to the relational, material, and interdependent contexts of the digital and networked home.

Chapter 7, "Displacements," considers how media technologies become obsolete, dysfunctional, and dispossessed and examines how households manage older technology devices and platforms and eventually move toward displacement, including through replacement and disposal. As new things are placed, older things are displaced; as new things are purposed, older things are repurposed; as older things deteriorate and break or become obsolete, they are repositioned within the ecology, or they are moved to the periphery of the household media ecology (storage), or they are moved from the house altogether to some distant place. All of this can be problematic. Items might not function but are retained for symbolic or sentimental values. Devices may be passed on to others, such as children, repurposed, and used in a niche or limited fashion within the wider domestic media ecology. Attachments are formed with material things that make disposal difficult, and the lingering functionality and use values of media continue to be exercised, challenging the imperative of consumer electronics to force obsolescence and upgrading. In turn, the chapter investigates the steady accumulation of unused media within the storage spaces of homes, informed by uncertainties around data stored on hard drives, electronic waste protocols, or even the slim chance of future reuse.

These household practices of sharing and storing technologies are contrasted with the ways some other legacy media technologies are ruthlessly disposed of, as exemplified by the large number of analog televisions littering suburban curbsides, evicted from the home but not the long tail of "zombie media" (Hertz and Parikka 2012) continuing to have an impact on other environments. Outside of the home, Chapter 7 situates these domestic practices of managing and discarding obsolete technologies within the wider political economy of digital media industries and discourses, in which technologies are continually overrun by newer technologies, and households are forced to engage not only with practical issues of obsolescence, legacy devices, and legacy media, and their configurations in ad hoc assemblages, but also with affective concerns around environmental consequences and the political economies of their electronic waste.

The book closes with a short conclusion that returns to the book's key concerns and themes; considers the broad range of ways in which people embrace digital media in their daily domestic lives; reflects on the ongoing changes in domestic media and communication technologies, platforms, and infrastructures; and addresses the broader implications of digital media materialities for contemporary household relations, economics, and environments.

[1]
HISTORIES

The primary concerns of this book are the ecology of communications technologies that have reshaped the domestic home over the last two decades, and these have been predominantly electronic—though electronic media have populated the home for much longer. Historically, the home's communications ecology was rearranged as a place was made for radio and gramophones in the 1890s, for telephones in the 1910s, television in the 1950s, video in the 1970s, home computing in the 1980s, and the internet in the 1990s. Each of these transformed what came before, often replacing the prominence or focality of one medium with another, though not necessarily obsolescing; each innovation shifted the functionality and location of older media to different spaces within the home, where they often persisted on the margins or in more niche roles. Each shift reshaped the home materially and, in Altheide's (1995) terms, altered its *domestic media ecology*—a term used throughout this book, and the focus of the next chapter, which designates the home as a media environment composed of dynamic interactions of people, devices, infrastructures, and data in situated relations.

As scholars like Carolyn Marvin have shown through their historical analyses of household media, the newness of the landline telephone initially established worryingly open lines of communication to the home and integrated the *social* and the *domestic* in ways that were novel, unsettling, and difficult to adjust to (Marvin 1988). As such technologies became ordinary and mundane in their domestic presence and use, other media took their place in terms of focality and uncertainty. Broadcasting, first via radio technology then television, served a new social need by bringing information and entertainment into the privatized home (Williams 1975)—indeed, bringing the world into the home. Lynn Spigel described the idea of TV bringing amusement into the living room and providing families with a window on the world as they received information and entertainment from outside the house as a form of privatized public entertainment (1992). However, the integration of such technologies was not necessarily smooth. The television required a great reconfiguration of the domestic environment, both materially and in its logic. Decisions had to be made about where this new technology belonged. With the advent of television in homes, the furniture industry started to promote sofas and chairs to maximize viewing comfort, and the setup of seats was now directed toward the television set instead of toward each other. Described as the "electronic hearth," the television started to replace the traditional hearth as a focal point of the living room (Tichi 1992); its function served as a social (family) medium, around which household members gathered. In addition, other objects in the room, such as pianos, had to make room for the television. Although the television materially and affectively replaced radio and piano before it, these technologies did not necessarily disappear but instead shifted to other (peripheral) spaces and uses. In turn, household relations had to be negotiated around such a focal entertainment medium, with relations of gender and power organized through viewing placement and control (Morley 1986).

Television thus changed the *material* environment and *social* practices of the home. Derham Grove's *TV Houses* (2004), for example, provides insight into the home cultures that co-emerged with the introduction of television to Australia in 1956. Those who could afford a television very often designated a room for television viewing—the "TV room"—and ancillary technologies were marketed for this new place. Television parties became common, giving new expression to gendered roles such as that of the wife as hostess. New packaged foods came on the market—foods that were quick to prepare, came in disposable foil, and could be easily consumed while watching television. With the television, the home truly became the center of consumption for much of the economy. This role remains today, and many houses still contain a room that might fairly be described as a TV room, though contemporary media consumers often use the term "home theatre" or "home entertainment center."

Even as television and its located uses were standardized, other media emerged to challenge its prominence. Consider the shift from mid-century collective family television viewing (Morley 1986), to the introduction of affordable and initially uncertain personal computers (Lally 2002), to the emergence of young people's media-rich bedroom cultures, and the increasing multiplication of screens enabled through mobile devices and the introduction of home internet and later wireless internet. While offering much more diversity in media entertainment, these changes have in turn been critiqued for fragmenting and individualizing family life through "living together separately" (Flichy 1995, cited in Livingstone 2002, 141). All of these waves of media have shaped fantasies and ideals of (mostly) middle-class life—such as the "smart home" as a particular model of digital domesticity that seamlessly integrates structure and computing technology (Spigel 2001).

In the early years of this century, when our empirical research on digital domesticity began, the domestic media ecology was populated with technologies that would be unfamiliar to most people today, and technologies that are ubiquitous today were absent. Most people did not have access to the internet from home, those who did generally used dial-up modems attached to landline telephones, and only the most enthusiastic had a faster ISDN[1] connection. Dial-up modems were sometimes plugged directly into the telephone-line socket, but, in older models, the telephone handset was attached to the modem via rubber suction cups. In either case, data transfer was painfully slow (by today's standards, though not by the standards of the day), and, importantly, the telephone and the internet could not be used at the same time. UseNet Newsgroups was a hugely popular site for online discussion. Static web pages were constructed with HTML editors and were published on the World Wide Web by businesses, universities, and some individuals. There were many such websites and Usenet discussion groups, but not so many as to prevent the publication of paper directories listing these sites and discussions. The internet had (controversially at the time) moved from a research, university, and community infrastructure to a media increasingly dominated by commercial entities, while, at the same time, the "information superhighway" was being promoted as a commerce-friendly alternative to the internet. Digital compact disks had displaced vinyl as the media of choice for music, and music was pirated via CD burners. Free-to-air television, movies, analog radio, and newspapers were the dominant mass media; streaming media as we now know it did not exist, though cable television was becoming popular in some locations, and some people without

---

1. Integrated services digital network, defined in the 1980s to carry speech and data on the same line but since decommissioned in most places and replaced by ADSL, cable, or fiber.

cable were eager enough to use dial-up to download pirated movies over, say, a 12-hour period. Personal desktop computers, such as the iMac and the IBM PC, were becoming popular, as were videogame consoles, and email was the killer app for digital communication. There were no smartphones, and very few people owned a mobile phone—exceptions being salespeople and tradespeople who installed a phone in their work vehicle, and well-to-do real estate agents. Some people used pagers to stay in contact (e.g., refrigerator mechanics and surgeons on call), but, of course, Facebook, Twitter, Instagram, text messaging, and other forms of social media did not yet exist.

Today, domestic life is increasingly characterized by connectivity and the ubiquitous use of communication and media technologies. Within the home, newer modes of communication, labor, socialization, and organization map onto and inflect traditional family practices. As social and technical sites, then, homes today have increasingly become critical nodal points in networked infrastructures connecting to the world via a raft of pipes and wires carrying water, energy, and, crucially, information (Graham and Marvin 2001). Homes are sites of production and reproduction, of labor and leisure, of parenting and socializing. Homes are sites where family members synchronize, come together, and play out the complex performances that constitute everyday family life. The home is still, in a sense, a refuge from the exigencies of employment and public life, yet it is now irrevocably enmeshed on a minute-by-minute basis with global events, wider society, and other places and times.

Yet, the home analyzed here is not a fixed entity; typically it is a building that both physically and socially changes over time. Architectural theorist Stewart Brand argues that the idea of residential architecture is permanence: "buildings loom over us and persist beyond us. They have the perfect memory of materiality . . . they are *designed* not to adapt" (Brand 1994, 2, emphasis in original). However, a diachronic understanding of the life of buildings emphasizes the process of adaptation. Against the modernist dictum "form follows function," which Brand criticizes as a misguided belief that function could be anticipated, he writes that "first we shape our buildings, then they shape us, then we shape them again—ad infinitum. Function reforms form, perpetually" (1994, 3). This functional melding of form plays out in the continual reshaping of homes in response to historical architectural norms, and to biographies of particular buildings. Brand asks: How does this process work? In the domestic context, homes respond "directly to the family's ideas and annoyances, growth and prospects. The house and its occupants mould to each other twenty-four hours, and the building accumulates the record of this intimacy" (5). It endures in time and remembers past inhabitants, and its internal dynamics adjust to the dwelling of inhabitants as much as inhabitants adjust to it. The process might be thought of as a cycling through imagination, construction, occupation, appropriation, domestication, dwelling, reimagining, reconstruction, reoccupation, and so on.

This process, however, occurs at different rates. Brand writes that "homes are the domain of slowly shifting fantasies and rapidly shifting needs" (6). He locates differing timescales through an expanded list of layers or components of a building outlined by architectural theorist Frank Duffy (1990). These are the *site* (geographic setting), *structure* (the building itself), *skin* (exterior surfaces), *services* (plumbing, wiring, etc.), *space-plan* (interior layout), and *stuff* (furniture, appliances, the movables or "mobilia"). Of particular interest to us in this book is the media and communications stuff. How do the inhabitants and their media stuff adapt to the physical spaces of a building's obstinacy and slower processes of change? And, how does the building respond to the faster changes and practices produced by the electronic media stuff within the home? And, finally, what about "the stuff of social practice" (Shove et al. 2007, 12)? How do the inhabitants of domestic spaces adopt and adapt to the ecologies of media stuff as it is domesticated, accumulated, and cajoled and ultimately fails to function and becomes obsolete?

In the section that follows, we give one vivid example from our work as a response to these questions.

THE HOME BRAIN, E-KEY, MERLIN, AND KEY4IT

On March 24, 1999, a commercial-in-confidence document was drafted by a group of Melbourne-based businesspeople setting out their ambitions for a technology they called the "Home Brain" (Stonehenge 1999b). The Home Brain was heroic in its ambitions, ambitions that retreated, expanded, and morphed through numerous design iterations as the technology was variously renamed "Merlin," the "e-Key," and finally "Key4IT." Each name change signified a different technological imaginary and informed a different sociotechnical ontology, and each was prescient in different ways of the technologies that have now become commonplace.

The Home Brain was composed of three interlocking systems:

> A Home Management System, which provided an interface to access from the home or from another networked location for systems such as security, heating, lighting, and sprinklers
> A Home Information System, which provided information to the homeowner about the building specifications, contracts, warranties, and service agreements that apply to a property and its contents

A Home Linking System connecting owners to their immediate neighborhood and community (Stonehenge 1999a).

Through these three systems, the Home Brain set out to be self-consciously futuristic—that is, it was enrolled to integrate the past and the future with Category 5 cable, 8 megabytes of RAM, a 500-megabyte hard drive, and a 33-MHz processor. First, the technology sought to provide the construction process with a history by tracing and recording significant moments, signposts, and decisions in the life of the house, from a comprehensive set of house plans, through details of all phases of construction, materials, and specifications, through all maintenance, all renovation, and eventual demolition. Second, the intention was to allow present technology to flow as smoothly as possible through to the future, insomuch as the developers anticipated that unanticipated demands and desires were certain in the medium or even short term. Excess capacity was therefore built in where possible, and open standards were adhered to in an attempt to "future proof" the development. Clichés such as "building tomorrow's house today" were never far from the tongue and pointed to this attempt at present/future integration. Third, the technology manifested the desire to integrate turn-of-the-millennium urban development with the coming century but also with an idealized vision of community from a past era. As present technologies that pointed to a future, they were asked to serve nostalgic purposes.

Two decades later, these goals for a smart home, a digital archive, and social networking might not seem ambitious, but the past was a different world:

> The Home Brain must [...] be something of a "wow" or status item for the home, similar in concept to modern home theatre

systems. It is the type of high-tech feature that, while at a party, the men will all stand around the monitor and talk about the "cool" things that the system can do (Boys and their toys syndrome). The placement and overall appearance of the home brain system must be something that the women do not see as unsightly, "in the way," or otherwise disagreeable. Rather, they should find it a useful, easy to use, beautiful and personal device. This "wow" factor is a function of the entire delivery system, interface and hardware design. (Stonehenge 1999b)

Meanwhile, the Home Brain's decedent, e-Key, promised residents the ability to achieve "total lifestyle management":

One of the most exciting aspects of the *e-Key* system, is its capacity to facilitate interaction between local residents to create a strong and harmonious community. Residents will communicate with neighbours, participate in special interests groups and form alliances with local traders, in ways never before realized. (Stonehenge 1999d)

One of the earliest sites of our research, Williams Bay, was positioned as the "first development of its kind in the world to pilot community development with 'life' changing technology. The benchmark for lifestyle development of the future" (Stonehenge 1999c).

A media ecology constituted by a digital mode for casual and organized interaction was in this way to be provided in what were at the time technically ideal conditions shared by all residents. Before any of this, though, for tens of thousands of years, the indigenous people who lived in Williams Bay called the area Koort Boork, which translates to "clump of she-oaks." About fifty people lived permanently on the peninsula—at that time a lightly wooded area of she-oaks, native cherry, and blackwood, giving way to saltbush sand dunes, and mudflats and mangroves nearer the water's edge. Koori graves and totemistic stones have been found near the former Williamstown Racecourse, but little trace is left of the original owners of the land. When the sheep were moved in, they were moved out.

The move of advanced digital communications systems from the workplace to the home, from the city to the suburbs, and from commerce to community was of course not limited to Australia. Well-known international examples paralleled and in some cases preceded the case studies we conducted. Examples are the Digitale Stad initiative in Amsterdam, Blacksburg in the United States (Silver 2000; Jankowski 2002, 42; Carrol 2003), Netville in Toronto (Hampton 2000; Hampton and Wellman 2000; Jankowski 2002), and the Japanese Multifunction Polis experiments (Inkster 1991).

AN UNSTABLE IMAGINARY

Like most technologies in their developmental stage, the Home Brain, Merlin, e-Key, and Key4IT took an ephemeral and unstable form as general plans and intentions, fragments of suggestions, prototypes, briefing notes, forum reports, formal design specifications, marketing documents, grant applications, imaginings, and back-of-the-envelope sketches. Through these forms the technology coalesced, broke apart, and reformed in fluid ways, configuring and reconfiguring a different entity with each change. In the face of this instability the developers had to speak for the form the technology was taking, both what it was and what it would do.

As the technology was imagined, each house would be equipped with its own server. This might be located in the study or under the stairs, or might be a virtual server at a remote location. Each server would be networked to other personal computers that members of the family might have, and to at least one port in each room—imagined as a touch-sensitive screen manufactured by Hewlett Packard, with a virtual or real keyboard. Smart home functionality was envisaged: each household system would interface with other electronic systems in the house (mobile phone, air conditioning, sprinklers, security systems, and so on), through Category 5 cabling or, more radically, through Bluetooth, a wireless technology developed just a year or two earlier by Ericsson. In turn, the household system would have a broad bandwidth connection to the fifty-two other servers in the Williams Bay development. Prescient of what today are called "walled suburbs," all fifty-two household systems would be protected and contained by a common portal and firewall, and the term local private network (LPN) was adopted to describe this network architecture. Beyond Williams Bay, the LPN would of course connect through to the internet.

At the level of the family, the technology was to provide a "virtual fridge door"—a message pad, a calendar and appointment book, a reminder service, a jobs-to-be-done list, and other asynchronous communications services, coordinating and integrating family members who share the same space but not necessarily the same time schedules (Figure 1.1). The Home Brain, Merlin, e-Key, and Key4IT were also to enroll and network the local community groups, clubs, and schools; enhance opportunities to work from home; offer opportunities for local businesses to network with the residents; and monitor and control electronic appliances such as heating, lighting, watering, or alarm systems from anywhere in the house or from anywhere with an internet connection.

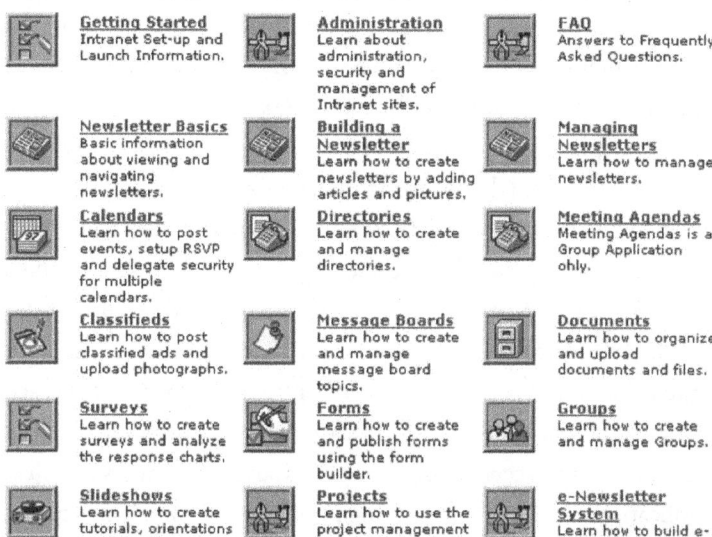

Figure 1.1. An early version of the Key4IT Help Page interface, indicating functions, 1995 (Wired Homes 2002–2004)

The imagined media ecology of Key4IT marked the urban environment with a series of concentric boundaries, each serving to define, defend, and admit. In the inner circle was each household's communication and control center, with its own server, interfaces, and networked devices and appliances to be accessed either physically or virtually by members of the household. Beyond the first circle lay the LPN encompassing the fifty-two Williams Bay households (and perhaps invited others), secured by a firewalled portal. The household boundary was permeated by default to allow the flow of certain data from the household to those within the second circle, and vice versa. The Williams Bay LPN boundary might be approached but not crossed by local businesses, service providers, community organizations, and the like without invitation from households, though they might paste messages, invitations, or requests. Certain trusted actors might be given the keys to the city gate, should they be accepted as contracted domestic service providers. Such actors would need access to the LPN and would need both physical access and data access to individual households in order to perform domestic maintenance tasks.

In terms of controlled data movement and data access, this technological ecology clearly paralleled the security-conscious walled suburbs or "gated communities" sometimes found in current-day North America, and some cities in Africa and the Middle East. Each system boundary set conditions for entry and egress and defined areas of differentiated privilege and power over the movement of information and access to information. These differentials had clear implications for further stratifying privilege, and for constructing a sharply differentiated ghetto or elite enclave rather than something that might be described as a community.

COMMUNITY

Amitai Etzioni (1995), Howard Rheingold (1995), and Robert D. Putnam (2000) were among several influential public commentators and intellectuals to use empirical and anecdotal evidence to suggest that by the turn of the current century people in developed Western nations were more isolated from one another, more distrustful of one another, and more fearful of the Other; more cynical about public service and public institutions, whether political, religious, educational, or corporate; more individualistic and less willing to act in the general good, for public benefit, out of duty, or noblesse oblige; and more likely to address aspirations in the private sphere rather than the public sphere—to look to the market to provide "for me and mine." In the 1980s in particular, the withdrawal of the public good as a target for social policy was speeded by a neoconservative, New Right, or economic rationalist ideological hegemony. The popular ethos over this time had been to increasingly demand private consumption, mediated through the market, for the satisfaction of personal rather than communal ideals or objectives (McLean and Voskresenskaya 1992). The public institutions and public utilities established in the last half of the nineteenth century and the first half of the twentieth century to provide education, power, health services, transport, communications, and so on were informed and constituted by a modernist discourse that was outmoded by the turn of the century. Digital technologies were of course deeply implicated in the construction of this changed ground deemed to be hostile to community: "No longer do *we*, as members of the group, belong to the community, rather the community belongs to *us*" (Jones 1997, original emphasis). We built our own social networks, and, within these networks, obligation and reciprocation coexist, often uneasily, with individualism—which remains the dominant mode of relations. However, though developments in information technologies may well have become part and parcel of the problem, there is no doubt that in many circles they were positioned by influential people such as Rheingold, as well as Stewart Brand, William Mitchell, Ray Kurzweil, John Gilmore, John Perry Barlow, Mitch Kapor, and many others, as part of the solution, if not *the* solution.

The rationale for building these new media ecologies brought together a mix of romantic communitarianism and modernist techno-utopianism, in significant respects motivated by a perceived loss of community and community values: "The gravest and most painful testimony of the modern world, the one that possibly involves all other testimonies to which this epoch must answer . . . is the testimony of the dissolution, the dislocation, or the conflagration of community" (Nancy 1991, 1). By the turn of this century hundreds of digital community networks and what we now call social networks were operating in North America, Europe, and Australia. Access to these networks was primarily from the home (and from work), and the home was in this sense the locus of a new form of community and new forms of social interaction. The media ecology of the home was perceptibly different, and the media ecology in which the home was situated was perceptibly different. While many of these networks were set up and run through the collaborative efforts of community organizations, residents' groups, local government authorities, corporate sponsors, university-based research groups, university IT faculties, and welfare and educational agencies, by 2005 entrepreneurs were exploring the commercial possibilities of community and social networks through vehicles such as GeoCities, the Globe.com, CompuServe, America Online, SixDegrees.com, Friendster, Yahoo, and many others.

Ideas that were important to the techno-utopian argument driving online social networking as a new and convivial media ecology were centered largely on theories of social capital. Digital social networks were rationalized in terms of strengthening ties between established friends, but also encouraging more extensive weak ties between people, bridging people with differences, norms of generalized reciprocity within social groups, and so on (boyd 2010). It was hoped that social media would go quite some way to build social capital and normalize reciprocity, with information sharing, favor trading, participation in voluntary groups, cooperation, providing assistance, and requesting assistance all becoming normalized.

In this new media ecology, the notion of a geographically based community, located where the home happened to be located, and constituted in recognition of common identity, interests, and obligations, needed to give way to a personal network as the new structure of sociality. In this construction, a social network was not a shared public good but a private asset, a personal store of social capital actively built and maintained by individuals to suit their own individual sense of identity, desires, needs, and interests. This representation of social life had little in common with traditional representations of community.

CONNECTED HOMES

By the early 2000s, online social networks were prominent in the media ecologies of homes. The accommodation of social media within the domestic media ecology was realizing, and in some cases normalizing, functionality that a few years earlier had provided the Home Brain with its "wow" factor. In the housing market, "community intranets," or web-based community networks, were becoming a more common part of the "lifestyle" package offered to purchasers of master-planned real estate developments in Australia. Mainstream developers, such as Delfin, Urban Pacific, Stonehenge, and the Docklands Authority, were installing community intranets in their new greenfield and highrise developments. Homes were connected electronically to the outside world less than a hundred years ago. And, in the current century, they have become a communications node in a global network, connecting to the local community; to work; to social, political, and commercial organizations; and to entertainment and service providers. This new role as an important site not only for the consumption of information but also for the production of information has had important implications for what the home is and does, and how families function within it.

The widespread use of the internet for social interaction thus posed a major challenge for parents and their children and constituted a new site for negotiating maturity, responsibility, freedoms, ethics, and risk (Livingstone 2002, 2009). One hundred years prior to this, the telephone disrupted family life and posed challenges to the status quo in similar ways (Marvin 1988). Using a telephone, a person could simply call, without introduction, and without visual and other contextual cues indicating identity. Like the internet in the first decade of this century, the telephone at the turn of the previous century was accompanied by a degree of moral panic about the outside world reaching into the privacy of the home. As has been the case with digital communications, new rules of propriety had to be developed around the telephone. On what grounds should one call, as opposed to visit or write? Should one answer one's own telephone or have someone field the call? If so, who should have this responsibility? Should one accept a call from a stranger? How should one deal with unwanted calls? Did it matter if one is dressed, or how one is dressed, when talking on the phone?

The dozens of families that we visited in the early years of this century had similar concerns. For example, parents needed to decide whether a child would be allowed to have internet access from a computer in the bedroom, where its use would be unsupervised, or, like the wired telephone, whether the internet should be used from one of the shared spaces in the household where supervision is implicit. If they decided that supervision was desirable, should this be a parental responsibility, or should it be automated and performed by "net-nanny" software? How much responsibility was to be passed to children to make their own decisions about internet use, and at what age should responsibility be passed? In an echo of Marvin's analysis of the landline telephone in the twentieth century, complete strangers from all over the world are entering the home via the internet and are communicating directly with the children of the household, raising all kinds of alarm bells for wary parents.

In addition to the perceived risks posed by uncontrolled communication, the internet also threatened ideals of good parenting on many other fronts. Children can and did stumble on inappropriate content, and can and did seek out explicit content, with fewer opportunities for parents to intervene. Violent images were entering the home in this online media ecology, and bullying began to be a feature of online interaction. Sexist and racist content was entering the home without parental control. "There is a dark side to the internet," one parent named Sam observed to us, "and an email address opens you up to that." Sam quipped that his daughter Olivia would have to wait until she turns twenty-one before she has her own email account, although he thought "twelve or thirteen is probably realistic" (Sam, Connected Homes 2004–2010). Whereas in previous eras iconic symbols of maturity involved giving children their own door key, obtaining a driver's license, or reaching the legal age to drink alcohol, in the latter part of the twentieth century it involved allowing children to have their own computer or email account and, in more recent times, their own mobile phone and social media accounts. We were often left with the impression that parents wished for less complicated domestic politics.

Each variation in the domestic media ecology, and each new point in their children's independence, were accompanied by complex family negotiations, while a common strategy was for parents to configure the material arrangements of media in the home. The domestic media ecology coexisted with many other domestic functions—cooking, cleaning, eating, talking, resting, working, sleeping, and so on—and there was ample room for conflict as the media ecology integrated with these other preexisting domestic regimes. Computer games of perceived dubious value were occupying hundreds of hours that young people might have been using for homework or for sports or hobbies.

In these early days of widespread domestic internet use, a consensus on parenting strategies was not evident to us. We found that parents had difficulty forming and maintaining a united front and often disagreed with one another. Decisions tended to be made on a case-by-case basis and were applied to a particular site or genre of sites, a particular app, a particular game, and particular times and particular modes of media use. Decisions were thoughtfully made and were not often ad hoc or arbitrary, but they were tentative and relatively uncertain, subject to renegotiation. Certain principles were commonly in evidence, relating to violence, sexism, communicating with strangers, divulging personal information, times and places in the house where technology can be used, and so on, but much internet use was ungoverned by these principles, in which case negotiations and decision making tended to be context specific and contingent. Even though decisions might have been contingent, negotiating acceptable and unacceptable internet use provided a new vehicle for parents to express a position on issues that went well beyond particular instances of internet use. Issues relating to relative maturity, facilitating growing independence, clarifying and expressing values and ethics, what friendship is, and clarifying and expressing what a parent is and should be and what a child is and should be were all played out through the domestic media ecology.

A good number of changes were evident in the performance of parenting in the period from 2000 to 2010. The increasing affordability of laptop computers and mobile phones, along with the introduction of broadband wireless systems and structural changes to the placement of digital technologies in the home, had afforded families the opportunity to use devices including the internet in locations throughout the house. In the first years of the twenty-first century, technology use was typically limited to particular places in the home; TV watching sometimes took place in the designated media room, and computers were used in the study or in the home office. By 2010, these families now worked in a range of spaces, including the bedrooms, dining rooms, and the hub of the house, the kitchen. Parents using laptops and wireless routers often worked in the kitchen space as it afforded them the opportunity to multitask (e.g., answer emails, make dinner, answer the phone) while still observing the children watching television and engaged in other activities. The purchase of large flat-screen TVs supported the traditional viewing practices of some of the families. As these technologies moved from dedicated and specialized locations, such as home offices or media rooms, out into the rest of the home, they were regarded as more ordinary and more closely integrated with other everyday domestic activities.

It was also apparent that views about the value of particular technologies changed over time as the technologies in question became less exotic. For example, some parents who expressed early concerns about children's use of MSN and the alien forms of socializing it enabled were, by the turn of the decade, encouraging their children to use new technologies (such as MSN) so that they could "fit in" socially. Ironically, this sometimes coincided with a change in attitude on the children's part; once a vital part of their social existence, MSN was longer exciting or indispensable, no longer essential to social life, and not an assertion of independence of adulthood. As it became mundane, MSN retained its utilitarian value but was no longer invested with the exotic.

The households we visited in this period were households that made use of the internet at a time when household internet use was the province of a privileged minority. Our families were, in this way, a self-selected sample of affluent, educated, professionally employed people. (Later, when digital technologies in the home became more common, our base of participating families became broader.) Among the sample were households that were "high-end" users of technology, families that were invested to some degree in the "technological sublime" (Nye 1996) and sought to realize these desires through the smart home. It is interesting to look back from a distance of years at how this home was configured in 2005, and how it was imagined, and at the themes that motivated the smart home then.

One such theme was a perceived need for security, for the home to pursue its traditional function as a safe stronghold in an uncertain world through contemporary technologies. Another theme, unsurprisingly, was the semi-automation of traditional household functions and housework, and another positioned the home as a sophisticated entertainment center. So, for example, smart homes, circa 2005, would have screens installed in all rooms, sometimes more than one per room, and also extending outdoors to BBQ areas, patios, and poolside. Some of these screens were tablet sized, large flat screens were 52 inches, and for cinema-style viewing an image was commonly projected onto a large retractable screen. For security purposes, these screens would be linked to multiple video cameras positioned strategically around the home and garden. In security mode, the screens would display images to anyone in the house, but could also be accessed remotely through the internet. In entertainment mode, screens displayed feed from FoxTel, DVDs, videos, digital photographs, and free-to-air television. All rooms were also equipped with wired speakers (in this pre-Bluetooth era) so that music could be distributed through the home. The children of the home commonly preferred to sit in one place and consume the entertainment, but parents valued being able to move around the house attending to work or to hobbies while the entertainment followed them. Having media distributed through the house provided a different orientation to space and place in the home. In the smart, media-saturated home of this era, householders were able to stay put and bring the media to them and were able to move without losing their connection to the media. While some screens and speakers are bigger than others, and better than others insofar as the distribution of technology was concerned, space within the home environment and media consumption were not sharply differentiated from other activities. Media were in every place and therefore no place in particular within the home.

The smart home was imagined to reduce work, but housework that was semi-automated, in 2005, comprised trivial tasks, such as switching lights on and off, opening and closing curtains, controlling heating settings, and controlling the entertainment systems, all of which was done via remote-control handsets, either from within the house or from without (Harper 2011) One family we visited had seven remote-control devices in a drawer in the living room, each with dozens of buttons, each controlling a different system, and collectively causing a great deal of confusion and frustration. Also frustrating was the need to call in a programmer and/or an electronics technician, sometimes at considerable expense, to maintain devices that might otherwise have been simple, such as light switches or door locks. In the tradition of the Home Brain, smart homes were "future proofed," such that kitchens were wired with "dark cable" for later use, and ports were provided in readiness for the fabled online refrigerator and other such devices, yet to arrive, even 17 years later.

It became common around this time for people in professional occupations to work from home. Though not nearly as widespread as the promises and threats made by authors such as Alvin Toffler (1980) and America Online's Steve Case, by the early part of this century working from home was clearly in evidence. A case that illustrates this point was the home of John and Mary; Mary was working as a publishing editor for an education body and John as a systems technician (Connected Homes 2004–2010). At this time of wired desktop computers and cable modems, paid and even unpaid work done in the home was generally executed in the study or home office.

Reflecting common views, Mary reported a preference for working at home—"the boss is not looking over my shoulder," "I don't have to do battle with the Eastern Freeway," "the boss's interruptions make it difficult to concentrate on the work at hand [at work]." She said the excellent technology at home "enables me to be excellent at my job and to deliver quality publications." Overall, she considered "the working environment [to be] better at home." But, though she preferred working from home to working at the office for reasons that were commonly held, Mary was also ambivalent about it, for reasons that were also commonly held (Gregg 2013). For example, concerns were raised about hours worked that were not claimed: "No one says I have to put in the time at home—it's more implicit than explicit—but they know that I'm not going to be able to meet particular deadlines unless I do it." Mary recognized that "the fact that we have cable internet, plus the appropriate desktop-publishing software, means that the boundaries between work and home become fuzzy." Like many others in these early days of "telework," she was concerned about erasing the line between work and home. On the one hand, "it is relatively liberating," "it's pretty nice," "it's a perk!" But, on the other hand, "It really annoys me—it impacts on family life and relationships, takes up too much time and interferes with my real life" (Connected Homes 2004–2010).

John, a technical assistant on call from home two weeks in every six, used their study when he was on call, accessing work either by internet on his own desktop computer, or by unplugging that from the modem and connecting his work-supplied laptop. During this two-week period, John had less control over the rhythms of the day than Mary did. John had to be available any time day or night, and he therefore took a mobile phone to bed. With the newly installed broadband, John noted, he could work from home almost as effectively as from work, enabled by automated systems that surveilled the performance of digital machines at work and sent machine-generated SMS and telephone calls to home if certain conditions were triggered. Like Mary, John was concerned about erasing the work–home boundary and, to maintain this boundary, never worked on the weekends and never checked his email from home (except when on call). John pointed out that, prior to installing the internet, he might still work from home, bringing home paperwork and problems in the head, but, with an internet connection, work was much more intrusive. Like almost one third of Australians in the workforce, John regularly worked from home, and he did so to "catch up" (ABS 2016), implying that worktime in the office was insufficient to perform the work required.

VISIONS OF BROADBAND INFRASTRUCTURE

By the second decade of this century, large-scale media ecologies in countries all over the world were changing radically, particularly through the proliferation of access to broadband, and through the popularity of mobile media access via smartphones. In Australia, this ecological change was most closely associated with the National Broadband Network (NBN). It was to deliver high-speed broadband to all Australian households, and thereby alter the domestic media ecology of all Australian households, using a combination of fiber-optic cable and wireless and satellite technologies (Figure 1.2).

The usual mix of techno-utopian promises accompanied the planning of this new digital ecology: enhanced communication through email, instant messaging, VoIP (voice over internet protocol services), and other communications services; time-saving activities, including telecommuting, online shopping, remote work and study opportunities, and information-gathering and -accessing services; price/product discovery; enhanced access to education and knowledge; better access to new online services such as social networking, media/entertainment, and professional services; substitution of physical services to services delivered electronically; and engagement in the online community.

Whether fiber-to-the-node or fiber-to-the-premises, the macro-environment within which the micro-environment of

Figure 1.2. NBN contractors installing fiber-optic cable (High-Speed Broadband 2011–2017)

the home sat was changing, along with expectations about the use and understanding of homes. These changes in the home included better opportunities for flexible management of work–life balance, expanded possibilities for home-based production and access to education and health, overcoming geographic barriers and the tyranny of distance for economic participation and productivity in regional areas, and reduced commuting and the associated impacts on congestion, fuel consumption, and carbon emissions on the environment (Access Economics 2010; Gregg and Wilson 2011).

Many of these impacts were premised on a shift in the functionality and architecture of houses, in parallel with the installation of broadband infrastructure, to transition the home into a platform that combined hardware and software to produce an environment of interoperable consumer electronics or automatic devices, based on networks of shared protocols and standards (e.g., Allon 2001; Trulove 2002; Venkatesh 1996). Moving beyond consumer electronics and automatic devices was an imagined home that was a platform for production as well as consumption (Apperley et al. 2011), returning the home to its medieval function as a place of work—albeit a very different kind of work.

Visions of broadband therefore foresaw a future where homes are increasingly hyper-connected and integrated, where limitations of time and distance in domestic and working life have been overcome. The infrastructuring of the home through broadband envisioned by NBN fit with wider projections to augment and expand on the existing network of media and communications technologies in the home, along with rhetoric that was and is shaped by ideas and imaginings of the home of the future (see Dourish and Bell 2011). As the authors of the Australian Centre for

Broadband Innovation (ACBI)'s Broadband Connected Homes study observe:

> [V]isions of the "home of the future" have often presented wildly optimistic scenarios, in which almost every household device is connected and automated to make life easier for people. Though technically possible, factors such as a lack of interoperability, high cost, or complexity, have meant the majority of homes are little closer to this vision than they were a decade ago. (ACBI 2012, 1)

Nonetheless, interconnectivity and integration of media technologies remained a key focus in visions of how the high-speed broadband home of the future would function. In its 2011 report "The Role and Potential of the NBN," the Australian government emphasized the need for a high-speed internet connection to allow simultaneous downloading of content across multiple devices. They wrote that "one of the biggest factors in the increasing demand for bandwidth is not just the speed required by the increasing number of advanced applications, but also the speed required to run many applications simultaneously" (Department of Broadband, Communications and the Digital Economy [DBCDE] 2011, 219). The report depicted a scenario in which a range of devices are drawing on single broadband connection in a household with speeds of 50 Mbps/20 Mbps.

As the number of internet-enabled media devices in the home continued to grow, high-speed broadband became an increasingly integral infrastructure in the functionality of homes. This gave rise to fresh visions of the smart home, in which multiple devices—the television, computer, mobile devices, home appliances, energy and water infrastructure, and security systems—were interlinked on a local network or through cloud computing and operated remotely through smartphone applications. According to the ACBI, "for the connected home to realise its full potential, it must evolve to become a platform" (ACBI 2012, 1). The smart home was often depicted as a highly networked, functional, and interconnected environment that responded organically to its inhabitants and featured a range of media devices synced across a single network. As such, high-speed broadband was forecast to become more seamlessly integrated into the physical architecture and underlying infrastructure of homes, as well as individual household technologies.

These claims were echoed in both the literature and in claims by governments about the future potential of the NBN. Alex Burns and Stephen McGrail (2012, 40) cautioned, however, that the NBN signaled "a return to the 1990s rhetoric of the internet as an 'information superhighway' in a new guise," and that predictions of a radical societal shift should be kept in perspective. In our report on broadband implementation in an inner suburb of Hobart (High-Speed Broadband 2011–2017), for example, we found speed was not the main attractor; rather, having constant access to it was: "For many households, speed is not the difference that makes a difference; rather it is the fact that broadband is ever-present and ever-available, whereas dial-up must be switched on and off" (Wilken et al. 2011, 5).

As such, sweeping visions of the future required more situated analyses of how households connected to, understood, and appropriated such technologies. A domestic media ecology, being what it is, a densely interconnected web of heterogeneous actors, means that the technical dimensions of provision and access to communication infrastructures, devices, or services are by themselves not enough to secure straightforward inclusion in the digital economy (e.g., Warschauer 2003; Mahar 2008; Park et al. 2013).

Mark Warschauer (2003), for example, noted that access and inclusion require a range of interconnected resources: physical (hardware/device), digital (connection), human (literacy), and social (social networks). As such, the concerns around access, which were traditionally grounded in debates around a "digital divide" and the presence or absence of an internet connection, shifted to consider questions about a "participation gap." Here, participation encompasses a range of technical, economic, and social resources, from speed and cost of services to forms of knowledge and expertise required to successfully understand and use technologies.

Australian homes are continually evolving technology environments. The trend toward an accumulation of devices, screens, and access in Australian homes is, for example, well documented by the Australian Communications and Media Authority (ACMA) and by the Australian Bureau of Statistics (ABS). In 2012 it was not uncommon for Australian family homes to have multiple televisions, computers, and mobile phones, while 79 percent of households had internet access and 92 percent of these connections were via broadband (DSL/ADSL; cable; mobile/wireless) (ABS 2014). Predictions of next-generation broadband anticipated, for example, enabling video streaming applications, high-quality videoconferencing, faster and larger file transfers, and a number of high-bandwidth applications to run simultaneously (Ezell et al. 2009; DBCDE 2010). In this context, the home would not just be a site for consumption or even supplemental work (Fenner and Renn 2004; Nansen et al. 2010) but instead would be transformed into a "remote office location." The digital home, then, was cast as an integral part of a network of digital living with seamless transitions between home, office, supermarket, school, and hospital. In the imaginary of the NBN, the digital home became a vital connection in the growing digital economy. Consequently, the ACMA described this household media ecology as technology rich. Contemporary homes, then, were media-saturated environments, and as sites of continual technological change the study of domestic appropriation of high-speed broadband required an approach that examined the domestic technological environment as a whole, or the media ecology of technologies and practices, not just a study of high-speed broadband in isolation (e.g., Shepherd et al. 2007; Wilken et al. 2011).

THE HOUSE AND THE NETWORK

Household media and broadband infrastructure are, of course, positioned in relation to the wider network as well as the domestic broadband infrastructure. Indeed, the location of the house and its relation to the wider network is a prerequisite for domestic broadband. Just as our participants were conscious of the placement and movement of media within their homes, so they were conscious of the home's position in relation to the network, though of course attitudes to this position differed. Some homeowners worried that their lack of fiber-optic connectivity would affect the price of their house. Some researched which residential areas already had fiber and chose a house in that area because of this access. Some noted that the position of their home had always been relevant to network technologies, where under ADSL technologies distance from the exchange determined the capacity of networked objects to interact. Still others decided to make use of the fiber connection to, in effect, geographically move their house (or at least, their IP address and interactive domestic space):

> We've got some fancy geoblocking stuff set up . . . so we can watch Netflix and stuff in the States without having to worry about that sort of stuff. . . . [T]here is very little in Australia that uses the bandwidth that we have. Little in terms of services. You have to go overseas if you want to take advantage of it or get any decent content or services . . . So, one of the perhaps more unlikely consequences of the NBN is that, for media consumption purposes, I live in America now. (Malcolm, High-Speed Broadband 2011–2017)

Broadband infrastructure thereby enabled and constrained the capacity of devices of many kinds to be mobile and to be positioned both beyond and within the home.

Over the course of our research, internal broadband infrastructure shifted from wiring between objects, to wireless communication, or where internal infrastructure was made to disappear by trying to hide it, as in this urban household: "There's one [speaker] hiding over there. I try to keep them relatively hidden, the cords" (Riley, High-Speed Broadband 2011–2017). Objects other than wires that are not part of the domestic internet infrastructure were also deliberately positioned or hidden, either for aesthetic or functional reasons: "We chose to have the [NBN] box placed where we could easily hide it with furniture" (Ashley, High-Speed Broadband 2011–2017). Sometimes whole devices were hidden, in this case a television: "Stephanie likes to pretend we don't have one. We have long debates about it. Stephanie thinks it is better to cover it" (Peter, High-Speed Broadband 2011–2017).

As part of the entangled ecologies of household media, then, are not only visible devices in the home but also their relation to the often invisible or hidden infrastructure of household internet, which work to help construct these environments as emergent on

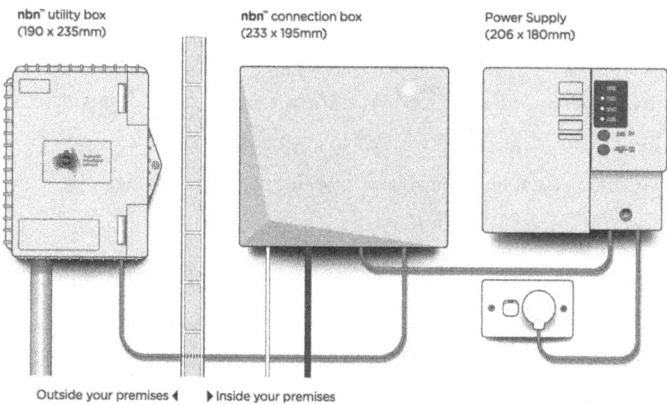

Figure 1.3. Diagram of NBN household connection hardware (High-Speed Broadband 2011–2017)

many scales of aggregation and disaggregation. So, for example, a wireless router connects to the power supply, the aerial, the circuit board, and so on. The wireless router is itself a component of the home network, which might itself be a component of a "smart home" communication and control system, which is in turn a component of the broadband network, and then the internet.

Thus, we found that the fiber-optic connection box was installed by householders in particular places based on options made available by the installers, by concerns about internet functionality, and by concerns about aesthetics. And, some of these concerns prioritized the relations between different media technologies, and object–object interactions. The connection box (an object positioned inside the house, sometimes called a NBN termination device (Figure 1.3) was often placed in a position that privileged its relations to other objects. Particularly important is its relations to the utility box (sometimes called the premises connection device), which sits outside the house and enables it to engage with the wider national network:

The box was put there because they asked, "Where do you want this?" The location is in relation to everyone else. Megan's room is over there. Dylan is down there. It's kind of central. It's also where I can see it. I also wanted to connect the office computer by Ethernet cable. Everything else is wireless. (Craig, High-Speed Broadband 2011–2017)

When they have troubles with the internet, which has happened a few times, Dennis is asked by the ISP to connect with Ethernet cable [directly to the NBN box], which justifies their decision to have it installed in the office rather than the garage [as it] is more easily accessible. (field notes, High-Speed Broadband 2011–2017)

MOBILITY OF MEDIA IN THE HOME

Changes in the national and the global media ecology have been paralleled by changes within the home, changes that might be described as sociotechnical in that they are associated with technical innovation and social innovation in ways that are impossible to disentangle. For example, through this century there has been a retreat within the home to ever more private places, aided and abetted by mobile devices and wireless technologies (Turkle 2017). The teenager's bedroom has been reinvented, and it is not uncommon for individual family members to have their own TV sets and music systems. Individual family members have their own internet-connected computers and iPads. We all have our own phones, and the days of having to take or make a call "in public," from a fixed phone in the hall, kitchen, or living room are long gone.

The contemporary home is a communications center and an important hub in global infrastructures of production and consumption. In this sense, the ecology of communications in the contemporary home makes it a part of the public sphere, the knowledge economy, and consumer culture. For many, the home is, again, a center of production. After a hundred years of separating, the work–home boundary is beginning to blur again.

The architecture of the contemporary home is also reflecting, accommodating, and driving this increasingly individuated media consumption and production, and it provides an environment that is a mix of premodern and postmodern. For example, the contemporary open-plan style of undifferentiated rooms without dividing walls, and without corridors, providing spaces for undifferentiated functions, is premodern. The open-plan kitchen/family room/living room is a multifunction polis for cooking, eating, entertaining, and working, and is a public/private space for family and nonfamily. But, ironically, within this space where family and others might gather, there is often very little communal socializing as the individuals gathered together attend to distant others through their devices, rather than to those occupying the same room. The fence and the front door symbolically and materially demarcate public space from private space, the home from the world, but, of course, contemporary technologies in the home and in the world pay no mind to doors or fences, and the boundary is erased in fundamental ways. The front door and the fence no longer work as intended.

A critical element of this change has been the infrastructuring of media homes by broadband internet connectivity, which is characterized by increasing bandwidth and wirelessness. Bandwidth and wirelessness have become part of a wider infrastructural understanding and imaginary of the digital home that has enabled the multiplication of devices, especially mobile and touchscreen devices, roaming easily within the home and transgressing the threshold of the door separating private and public worlds (Mackenzie 2010).

These dense ecologies of media are creating environments where media screens, connection, and communication are persistent and ambient within contemporary homes. We have seen, over the last seventeen years, a steady accumulation of media technologies, with each wave of innovation adding to the existing suite of media in the home. In particular, since the launch of the Apple iPhone in 2007 and the Apple iPad in 2010, and the myriad mobile devices that have followed in their wake, household media environments are increasingly characterized by the presence of multiple and mobile touchscreen devices, including a range of other touchscreen smartphones and table manufacturers along with the dominant Apple brand. The purchase of multiple devices, and the ambient presence and circulation of mobile media, was discussed by families in terms of the rationale for and patterns of household mobile media adoption, which often revolved around the perceived needs or benefits for children, articulated in terms of their suitability for use by children. Tablet devices, in particular, were perceived to offer affordances for children associated with usability, mobility, and multifunctionality, while key events, such as a planned family vacation, were often the catalyst for purchase. Once purchased, family discussions highlighted how mobile devices prompted more frequent, distributed, and varied use. The materialities of mobile media, in which they are connected to the internet, are not often located in a fixed place but circulate around the home, and are responsive to touch, meant devices were often lying about, available, and, enticing to use.

Such mobile media technologies are, however, not necessarily replacing or supplanting older media and communications devices, but they are having an impact on traditional telecommunications technologies. Increasingly relying on newer "alternative communications technologies, most commonly going 'mobile only' and/or VoIP [e.g., Skype] via their computer" (ACMA 2011b, 8), we see patterns of diffusion that push other media to the periphery or remediate, and displace, their traditional form. Services such as Apple TV now allow content to be streamed to a range of devices via the internet, meaning that video content is less and less tethered to the television set or the schedule dictated by television programmers. Likewise, radio persists, though it is remediated by the use of digital radio platforms and the rise of podcasting. Rather than substituting for older media, however, the addition of newer devices is being used to complement traditional activities such as television viewing with other so-called second-screen activities, such as simultaneous internet browsing and social networking. While multitasking may not necessarily be new, certainly the increasing speed of the internet alongside the mobility of devices is contributing to a situation that affords new configurations, patterns, and intensities of multiple-screen activities (see Highmore 2010).

As well as the myriad networked (and non-networked) media devices in the typical household, a number of trends are continuing to shape the way people use and consume media content at home. A significant development in the digital media landscape over the past few years has been the rise of social networking, as well as the growing number of individuals creating their own media content. Another important development that has changed media consumption and access in the home has been cloud computing—the ability to store information "in the cloud" on a remote server over the internet, rather than on one's own computer or device. As ACMA (2011b) noted, cloud computing has long been an important part of everyday internet use, from email services like Gmail and Hotmail storing users' emails online, to content-sharing sites like Flickr and

YouTube. However, cloud computing has also become increasingly important for the interconnectivity of media devices in the home. The ACBI noted that cloud computing "allows multiple devices in the home to share common data, without necessarily being able to communicate between each other . . . having a common software platform across a range of devices allows multiple devices to be part of a service or set of services that would otherwise not be possible" (2012, 27–28). For instance, having a Dropbox account allows families and friends to share documents, images, and video across devices without the need to physically transport it with a USB stick, while services like Apple's iCloud allows users to download music from the cloud as they need it rather than storing it locally. However, this has meant that, increasingly, homes have come to rely on online access for media content; when the internet connection goes down, so does access to a lot of media content like music, video, and gaming as these devices are severed from their proprietary services. As a result, a persistent internet connection is increasingly vital for enjoying the full range of media technologies in the home.

Such accumulation of newer mobile media in the 2010s has in turn placed increased attention on household economies of internet and communication, with families balancing the growing demands on bandwidth with the costs of upgrading internet speeds or plans:

> Before, when we were on the old plan, it was amazing how much data things like YouTube used. We were busting through our cap constantly, and getting shaped. It was an appalling system, it was horrible. You could almost never escape it unless you turned the thing off for a few weeks to let it die down again. 'Cause once you hit that cap, it basically stopped you using it, it shaped you back

to basically dial-up speeds. (Howard, High-Speed Broadband 2011–2017)

I reckon in the next few years we'll probably get rid of it [landline telephone], I don't know. It just depends on what we will replace it with, it's just a question of what you replace it with. In the near future it'll probably get replaced with something else, but as to whether it's mobiles or voice over IP or something else I don't know what that would be, but yes I imagine it will go . . . both of us have got mobiles, so that would be the logical step. I just don't make a whole lot of calls. You'd need to get a plan that would then replace a landline and a mobile at the same time is not cost effective. (Simon, High-Speed Broadband 2011–2017)

Finally, such ecologies have undergone dramatic shifts through developments in user interfaces, with input devices such as keyboards and mice amended by newer forms of touchscreen interfaces. Such media interfaces have a number of implications of household media use, enabling more embodied and gestural modes of interaction—what often falls under the paradigm of technology interaction known as natural user interfaces (NUIs)—that are celebrated for inaugurating a more intuitive or natural mode of interfacing with computation and that extend possibilities of use to more embodied and casual forms of use, while lowering thresholds of usability to ever-younger populations, with the so-called minimum user competency of touchscreen interfaces beginning from around the age of 12 months (Cristia and Seidl 2015; Hourcade et al. 2015). Touchscreen media, then, come to inhabit homes in ways that become readily available but also appealing to use, requiring less attention or effort, or through the affordances of the interface responding to touch.

So, it is not only the material structure of the home that is being impacted by high-speed broadband; as media devices become increasingly haptic and tactile, it is predicted that more embodied forms of household media use will emerge. Ruthven (2012, 34) has noted, for example, that "three-dimensional images are already in use in cinemas, and 3D-enabled TVs are on the market":

> Over the next decade, we anticipate that this technology will mature and become a standard feature of TVs and monitors, thus enabling even more life-like virtual interactions between people. With the availability of high-speed broadband networks, the use of holographic images may also become more commonplace. (Ruthven 2012, 34)

As well as broadband making virtual communication more widespread, Ruthven also predicts that it will aid the merging of media devices with the body itself, as "surface technology—including the human skin—will be a platform for accessing information" (2012, 34). He notes that "current science fiction themes such as cybernetics and augmented human intelligence, including the direct communication between the human brain and the world's data network, may become reality. These applications will require superfast networks to facilitate the seamless interaction between humans and their environment" (2012, 35).

Despite such diffusion of internet infrastructure and mobile media technologies implying a sense of inevitability in configuring ecologies of continual and consistent connection, along with visions of near futures in which homes evolve into seamless nodes of interactivity and productivity, supported by smart home devices such as voice assistants, as we have found and detail in the next sections, media appropriation, management, and negotiation is a much more contextual and complex process, which is neither uniform nor predictable. To approach such entangled temporal and spatial dynamics of digital domesticity, we turn to theoretical perspectives around ecologies in the following chapter.

[2]

ECOLOGIES

Since the turn of this century, households have undergone a dramatic transformation in the number, variety, and intensity of digital media and communication technologies, and an associated transformation in the way people use and interact with them. "Our domestic life," according to Roger Silverstone and Eric Hirsch (1992, 1), is "suffused by technology, and information and communication technologies are becoming a central component of family and household culture." This suffusion of technology has led to the domestic home taking on manifold functions, where it serves as a place of leisure, a command and control center, a place for production, and a place for consumption.

As a place of leisure, the domestic home, infused with media and technology, promises us leisure context and leisure content. In terms of context, we are promised more time for ourselves (to be delivered through online, automated, and labor-saving services), creature comforts (to be delivered through automated, "smart home" technologies), and a sense of security (to be delivered through intelligent surveillance devices). In terms of content, digital domesticity implies a bewildering array of integrated, interactive home entertainment devices, for music production, television viewing, interactive game playing, web browsing, chatting with distant others, exchanging files, and streaming televisual content. In addition, the materiality of broadband, digital television, home theaters, and the like manifests a broad cultural shift that relocated public entertainment and public spectacle from shared public/private spaces, such as cinemas, to private spaces, first in the home living room, now dispersed to bedrooms and other locations in the home.

As a command and control center, these domesticated media promise us access to detailed, real-time information about our finances, our consumption patterns, our commitments and priorities, and one another, with the potential implied for increased control of our finances, consumption, commitments and priorities, and one another. As a center of production, the contemporary digital home trends back to its premodern role at the center of human activity. As a center of contemporary production, the home accumulates, produces, and transmits information in vast quantities. It does this in four ways. First, the home is, like a multitude of other physical and digital sites, a data mine of considerable value, exploited through invisible, routine surveillance and processing of commercial transactions and home utility and service use (Zuboff 2019). Second, the home is used as a communications and publishing center by members of the household, in the course of coordination, recreation, socializing, and self-expression. Third, as the boundaries that confine work to defined places and times have weakened and disappeared, and the home returns to its role in the market economy, it takes its place as a data-processing center (Gregg 2013). Fourth, to shift registers, digital domesticity, as we discuss in Chapter 4, also produces the subjects in the home. Media and communications technologies inscribe representations of users in their design, and by attributing and delegating a variety of responsibilities, competencies, needs, and desires (Oudshoorn et al. 2004); these technologies interpolate users of various kinds.

Like traditional domestic technologies such as "white goods" (e.g., washing machines), and "brown goods" (e.g., hi-fi systems) (Cockburn and Ormrod 1993), contemporary media are not just objects of consumption but are also productive of genders, norms, and roles. That is to say, they inscribe values apart from those associated with functionality and usefulness (Gaver and Martin 2000).

Alongside production, contemporary digital domesticity is also significant as a center of media and technology consumption. Historically, information and communication technologies migrated from their place of origin—the workplace—to the domestic environment. Telephones, fax machines, mobile phones, computers, printers, and scanners were all initially designed and marketed with the workplace in mind, only later to find their way into homes. This is, of course, a process that is ongoing. In addition to transferral, we also see that the home itself is often the first and primary target for media and technology innovation and marketing (Venkatesh 1996). Products such as interactive digital television services; home shopping and home banking; "smart home" control of security, utilities, appliances, and services; integrated, computer-based entertainment hubs; and "intelligent," networked appliances have all been marketed principally to the home rather than the workplace (Harper 2011).

And so, as a place of leisure, a command and control center, a place of production, and a place of consumption, the home may now take its place as a fully integrated and articulated node in the digital space of flows. But what are we to make of this contemporary configuration of the home in order to be better entertained, organized, reachable, productive, consuming, and communicative at all times? What do we make of the effects and affects of these increasingly powerful, insistent, and capable actors that now inhabit the most private of our environments? How, in short, do we account, theoretically and methodologically, for households and the shifting role of technology within them—particularly given the dynamism of the former, and the complexification of the latter?

Responding to these theoretical and methodological questions is the concern of this chapter. In this chapter, we argue that the pioneering domestication approach to uncovering "the ways in which the cultural and social spaces of the home and social dynamics of the household are reconfigured by media and information technologies" (Chambers 2016, 45) has long been and continues to be an important critical approach. While the domestication approach remains influential and significant, we argue that it, alone, does not account sufficiently for the full complexity and fluidity of technological change within the digital home, and that additional theoretical models and methodological innovations and approaches have been and are required. In our own work, we have found that communicative and media ecology approaches, as developed by the likes of David Altheide, Matthew Fuller, and others, provide a productive additional theoretical framework—or layer—for making critical sense of the complications and dense set of interactions that give shape to the connected home. We have also developed a unique critical toolbox, the "domestic probe," for tracing and recording household interactions with multiple (and multiplying) connected devices and for capturing ecosystem complexity.

The chapter begins by revisiting the domestication approach, examining the important contributions this approach has made to facilitating deepened understanding of households and home technology use, as well as more recent applications of, and adjustments to, this approach. In the second part of the chapter, we then mount an argument for why ecological approaches form an important complement to established domestication approaches in accounting for even more intense forms of device take-up, use, and non-use within the contemporary home. And, in the third and final part of the chapter, we detail the development and implementation of the domestic probe as methodological means of capturing domestic ecology complexity.

THE DOMESTICATION APPROACH

The domestication approach offers both a theoretical lens and a methodological approach. Emerging from material anthropology and the sociology of consumption, it has been adapted and applied in media and technology studies to describe the processes by which technologies are integrated or domesticated into everyday life, and by the processes in which users and environments change and adapt accordingly. In this view, the domestic space is conceived of not in terms of static limits but as a site or locus in which devices are "'tamed' in different ways at different times, reflecting both technological and personal change" (Haddon 2011, 317), a process by which technology gradually moves into and is naturalized within domestic space.

The domestication approach considers both the practical and the symbolic aspects of the adoption and use of technologies, showing how these two elements, the meanings of things and their materiality, are intertwined and become part of everyday life. It aims to get at these things by tracing various stages—appropriation, objectification, incorporation, and conversion (Silverstone, Hirsch and Morley 1992)—in the process of acquiring, accommodating, using, and understanding what were at the time new technologies (e.g., Silverstone and Hirsch 1992; Silverstone and Haddon 1996). *Appropriation* is the process of purchasing a commodity and then bringing it into a household. *Objectification* is the process of physically and symbolically placing objects in the spatial environment of the home, and thus also the construction of the environment; technologies here have a functional and aesthetic significance, based on what they do and their appearance. *Incorporation* concerns the ways that technologies are used, and includes both their designed or intended uses, as well as unforeseen or alternative uses. And *conversion* defines the relationship between the household and the outside world; conversion is the remaking of the meanings or values and norms associated with the technology and the transfer of these back to the "outside" world.

In the case of the television, for example, this approach could be applied to understand how this technology was *appropriated*, bought as a consumer item off the shelf, as a functional product but also as an expression of social values and ideas related to private comfort and entertainment. The *objectification* of the television might involve finding a physical place for it in the house, where it is displayed (as status object and for viewing content), and then the environment around it adapted to accommodate its presence. Through *incorporation*, the television is used, fitted in with domestic routines and practices, the everyday life of households, integrating them into the daily patterns, structures, and values of the household. In the case of television, this was strongly related to the sustaining of domestic routines through the broadcast schedules. And, through *conversion*, it is possible to trace how the television connects the household with the world of shared meanings and claims of status and belonging. This conversion demonstrates the importance of the need of households to legitimate their participation in shared culture, especially broadcast and consumer culture (see Williams 1975).

The vocabulary and orientation of the domestication approach have been employed to consider both the practical and the symbolic aspects of the adoption and use of technologies, showing how these two elements—the meanings of things *and* their materiality—are intertwined and become part of everyday domestic life in which environments and users change and adapt. Thus, studies of "media homes" informed by domestication approaches were established over the course of the late twentieth century to discuss the ways that different waves of media technologies, such as televisions and computers, became physically and symbolically located within home spaces (Silverstone and Hirsch 1992; Silverstone and Haddon 1996). This included the shift from mid-century collective family television viewing (Morley 1980, 1986), to the introduction of affordable home computers (Haddon 1992; Murdock, Hartmann, and Gray 1992; Wheelock 1992; Lally 2002), the internet (Bergman and van Zoonen 1999; Ward 2005), and mobile phones (Haddon 2003; Morley 2003). Domestication-related studies are also concerned with how these technologies influenced household relations and imaginaries—from young people's media-rich bedroom culture, and the increasing multiplication of screens that have been critiqued for fragmenting and individualizing family life through "living together separately" (Flichy 1995; Livingstone 2002), to fantasies and ideals of middle-class life, such as the "smart home" as a model of digital domesticity that seamlessly integrates structure and computing technology (Spigel 2001; Harper 2011), to ideas about the "dislocation of domesticity" or the "domestication of elsewhere" through personalized and mobile electronic media (Morley 2003).

These studies share a common thread of addressing the changing character of lived relations (work, leisure, consumption, etc.) that result from the affordances, proliferation, and accommodation of an ever-increasing abundance of media technologies across the latter half of the twentieth century. And they point to mutual and continual processes of adjustment for practices, spaces (Lally 2002; Baillie and Benyon 2008), and rhythms of home life (Silverstone 1993; Nansen et al. 2009).

In this way, the domestication approach has proven helpful in addressing the unidirectional, universal, and decontextualized limitations of diffusion models of adoption that tend to be centered on broad social trends of technology adoption. Diffusion models of technology adoption are amended and complemented by the domestication approach insofar as the latter develops a dynamic means of assessing the more situated processes of adoption by going beyond individual motivations or single technologies and addressing the interplay of meaning and material affordances in the recursive interaction of people and technologies.

DOMESTICATION AND THE DIGITAL HOME

An important question to ask at this juncture is: How does the domestication approach help us to understand the rapidly accelerating processes of technology adoption and appropriation associated with the contemporary home? The domestication approach remains of enduring significance, both within our own work and elsewhere, for understanding processes of technology adoption and use (see Berker et al. 2006a; Chambers 2016, 52–61). However, it is also clear that adaptations and adjustments are needed if it is to remain relevant. David Morley, one of the key figures in the development of the domestication approach, has pondered this issue at length. Reflecting on the porousness of household boundaries and the increasing diffusion of technologies once thought of as exclusively domestic (the television, the phone), Morley (2007) suggests that domestication approaches need to give consideration to the "dis-location" of these technologies. Just as television "escaped" from the home to "colonise public space," "individualised media, such as the mobile phone, now contribute to the radical dislocation of domesticity" (Morley 2007, 7; see also McCarthy 2001; Bolin 2004; Lim 2016). Morley also acknowledges that, in a contemporary context, domestication approaches must move beyond a focus on single technologies in order to account for the radical reconfigurations and intensifications of what Hermann Bausinger (1984, 349) once referred to as domestic "media ensembles." Morley (2007, 200) calls for "non-mediacentric form[s] of media studies" (see Moores 2005; Morley 2009; Krajina, Moores, and Morley 2014;) in order to "understand the variety of ways in which new and old media accommodate each other and coexist in symbiotic forms" (Morley 2007, 200). Elsewhere, others have also called for revised models of domestication that account, not just for the arrival and domestication of media and communication technologies but also for their "re- and de-domestication" (Berker et al. 2006b, 3)—that is to say, the ways in which these technologies adapt and morph, are repurposed, or exit the home through processes of divestment (see McCracken 1988; Gregson 2007; Miller 2012), concerns that we take up and address in later chapters.

BEYOND DOMESTICATION

In this book, we both build on and depart from domestication approaches within media studies. We do so both theoretically and methodologically, by viewing technology use through a wider range of disciplinary lenses and approaches, moving beyond the symbolic significance of media and media content to account for the material presence of media technologies, their relations to householders and to other technologies, and their performance in shaping the contemporary digital home.

If we look beyond media studies, we see research on household media that has figured in other social, technical, and design-related literatures and that expresses a renewed interest in the material culture of the home. Our research contributes to this and draws on theoretical insights from this work in material anthropology, material culture studies, and cultural geography (e.g., Cziksentmihaltyi and Rochberg-Halton 1981; Miller 2001; Pink 2004; Blunt 2005; Blunt and Dowling 2006; Gregson 2007), each of which, in their various ways, expands the focus on domestic media to consider broader— and often mundane—material cultures of home life. This literature addresses habitation and co-habitation through empirical studies of the practices, habits, and routines of dwelling. For example, being "at home" is realized through "living with things"—that is, the acquisition, display, storage, and disposal of objects (Gregson 2007); through the meaning and symbolic value actually attributed by householders to artifacts and categories of household objects (Cziksentmihaltyi and Rochberg-Halton 1981; Miller 2009); through performed interactions with non-humans such as pets and gardens (Hitchings 2004; Franklin 2006); through the ways families collectively appropriate artifacts in order to "design" everyday household systems (Wakkary and Maestri 2008); or through the implications of architectural structures themselves, in which the performance of domestic life is also shaped in relation to the structure, design, obstinance, and imaginaries associated with the house and home (Rybczynski 1986; Spigel 1992; Brand 1994; Spigel 2001; Shove 2003; Hand, Shove, and Southerton 2007; Shove et al. 2007). These are issues that are addressed directly in the following chapter.

These socio-technical and socio-material studies of lived relations and dynamics within the home have also informed recent ideas of material ecologies and the mapping of entanglements and arrangements of people, technologies, practices, and spaces (Crabtree and Rodden 2004; Hearn and Foth 2007; Shepherd et al. 2007). Further, this literature has been reinforced by a "material turn" in the social sciences and in humanities housing studies. Adopting idioms from interdisciplines such as science and technology studies (see, for example, Hommels 2005) and actor-network theory (see Law and Hassard 1999), this material turn moves against the legacy of poststructuralist concerns and the dominance of semiotic approaches to social and domestic life (discursively ordered through symbolic human meanings of home) and instead addresses "non-representational" (Thrift 2007) theories and approaches. These theories and approaches account for the non-human (Latour 1993), the post-social (Knorr-Cetina 1997), the post-human (Hayles 1999), object-oriented ontologies (Arnold et al. 2016; Harman 2018), and the quasi-object (Serres 1982), which are approaches that dissolve binaries, recognize hybrids, are non-anthropocentric, and acknowledge distributed and evolving ontologies. This aggregation calls for and puts into practice a reinvigorated and symmetrical approach to the study of home life through analysis of the physical encounters and cultures formed within the home—the entanglements of spaces, objects, and subjects—in what Alison Blunt (2005) describes as domestic "cohabitation."

MEDIA AND COMMUNICATION ECOLOGIES

While the above theoretical approaches have informed the arguments of this book in significant ways, it is work more specifically on media and communication ecologies that has proved most influential in providing us with the necessary conceptual tools for making critical sense of the dense interrelations of multiple technologies and multiple people that configure the contemporary digital home (see Shepherd et al. 2007; Nansen et al. 2009, 2011; Wilken et al. 2014). These methods are tuned not to this or that media technology but to the entire domestic ecology that media technologies and householders collectively create. Our curiosity is about how these ecologies are shaped and experienced by the people and the objects that act within them through time, in a dynamic and ongoing way.

The metaphor of ecology was introduced into studies of media and communication in the 1960s and 1970s by theorists Marshall McLuhan and Neil Postman, who adapted it from already established ideas of "human ecology" in urban sociology from the Chicago School, rather than directly from biological science (Heise 2002; Strate 2004). The metaphor of ecology shifted the attention on communication media from metaphors associated with conduits for transmission or languages of representation (Meyrowitz 1993), opening a perspective onto the interconnectedness of media technologies configuring and stabilizing a cultural environment. Yet, in its metaphorical application, as Steward Pickett and Mary Cadenasso (2002, 6) note, the concept of an ecological system "can be used to stand for equilibrium, resistance or resilience, diversity, and adaptability." It is the first of these senses—that of equilibrium—that tends to characterize established, predominantly North American–based scholarship on media ecology (e.g., Strate 2004, 2006; Scolari 2012; Cali 2017; Strate 2017). For instance, Postman explains:

> We put the word "media" in the front of the word "ecology" to suggest that we were not simply interested in media, but in the

ways in which the interaction between media and human beings gives a culture its character and, one might say, helps a culture to maintain symbolic balance. (quoted in Strate 2004, 3)

We share Postman's focus on interactions, as well as the view that "the introduction of any new agent into an environment changes that environment" (Cali 2017, 10; see also Scolari 2012). What we don't share, however, is the strongly humanist approach (Postman 2000; Strate 2006, 92), the focus on "transformations of human consciousness" (Cali 2017, 17–23), and rather prescriptive and restrictive policing of terminology ("this other usage of *media ecology* is based on common sense definitions of *media*, rather than the broader understandings of *medium* associated with media ecology"—Strate 2017, 8, emphasis in original) that are associated with this strand of writing on media ecology.

Rather, informing the investigations of this book are alternative understandings of and work on media ecology, such as David Altheide's (1995) work on communicative ecology and the treatment given by media anthropologist Jo Tacchi (2006) and by media theorist Matthew Fuller (2005).

Altheide draws out four crucial aspects of the term "ecology" for the study of media, information, and communication technologies—formerly referred to as ICTs—which he summarizes as follows. First, the word "implies relationships related through process and interaction" (Altheide 1995, 10), relationships, what's more, that foster "interdependence, mutuality and co-existence" (10) and "opportunities for successful adaptations" (Nardi and O'Day 1999, 96). Second, it implies "a spatial and relational basis for a subject matter" (Altheide 1995, 10), meaning that "the characteristics of a medium depend on a certain arrangement of elements" (10). Third, "the relations are not haphazard or wholly arbitrary"

(10). Fourth, "there are developmental, contingent, and emergent features of ecology," which suggests that "ecology does not exist as a thing, but is a fluid structure" (11). This fourfold understanding of ecology informs his concept of "ecology of communication." It is a framework that is productive for grasping "how social activities are joined interactively in a communication environment [and] with information technology" (2).

Tacchi's work complements the above understandings and extends them in one crucial respect: by drawing attention to the macro-level "environmental" frames within which domestic and local communication ecologies operate. "It is a characteristic of anthropological media research," she writes, "that it considers media in wider contexts" (Tacchi 2006, 5). Drawing on extensive fieldwork in the area of media for development, Tacchi argues that the communicative ecology framework is useful insofar as it focuses attention not just on more immediate communication-related aspects of the contexts in which people operate but also on the ways that they are "in turn imbricated in other structural, social, economic and cultural contexts" (Tacchi 2006, 5).

Meanwhile, Fuller (2007, 2) writes that the term "ecology" is useful in that "it is one of the most expressive [that] language currently has to indicate the massive and dynamic interrelation of processes and objects, beings and things, patterns and matter." Emphasizing the dynamic nature of these ecological interactions, Fuller (2007, 1) argues that "complex objects such as media systems" should be understood as involving "processes embodied as objects, as elements in a composition" that settle "temporarily into what passes for a stable state" before reforming and resettling, and so on, in a process that is ongoing.

Finnish media theorist Jussi Parikka develops a similar line of argument to Fuller. Drawing on the work of Bruno Latour (2002),

Parikka (2011a, n.p.) suggests that media and communication technologies ought to be understood as part of "an environment of relations in which time, space and agency emerge." Thus, he argues, media are "less a matter of mediation and communication between humans, than a milieu of engagement, or relationality for the objects, vectors, agencies and processes that enter into its sphere" (Parikka 2011a, n.p.).

These insights into the dynamism and relationality of human and non-human actors within media or communicative ecological systems are helpful for making sense of the variety of factors and forces that impact and shape heavily "infomated" households (Darrah, English-Lueck, and Saveri 1997)—households that have and interact with a plethora of connected devices that collectively form "ecosystems of technology" (English-Lueck 1998, 6).

ENTERING THE DIGITAL HOME: RESEARCH CHALLENGES AND METHODOLOGICAL INNOVATIONS

As we have seen in this chapter and earlier in this book, the home is an historically fluid and continually evolving institution—at times a sanctuary, meeting place, private place, launching place, reproductive space, material asset, headquarters, resource base, resting place, entertainment center, and so on for changing combinations of people. Focusing on the home's performance as a node in a network of connections, and as a dynamic communicative ecosystem, enables us to focus on that fluidity. That is to say, this focus enables us to explore how the home is reshaped or reconfigured socially and materially to accommodate multiple technologies; how boundaries are drawn around the home and how it is patrolled to control entry and egress in a space of flows; how the technologies are distributed within the home; which sociotechnical "spaces" are private and which are communal; who uses them and who does not, when, and for what, and under what circumstances; how technologies perform to interpolate the home and the homemakers; how technologies are matched to particular purposes in the home; how the technologies interact, compare, complement, and compete with one another; who the communicators, and who the technophiles and technophobes, are; and how technologies are appropriated, domesticated, and ultimately discarded.

A key research aim, then, has been to gain access to households in order to develop a better understanding of digital domesticity and the "'intimate histories' of how we live with a variety of media" and other technologies (Morley 2007, 204) through examining these technologies in use, in situ. This, however, is more easily said than done. A major hurdle facing us and other researchers is that the home is, of course, a quintessentially private space—one that comprises "narratives, practices and sensory experiences that [are] not usually available for public view" (Pink 2004, 1)—and its peculiar and important character flows from this. Direct observation, particularly of an extended nature, is not always possible or desirable in private settings. Participant observation, for example, is a reliable and justifiably well-regarded ethnographic method for application in the workplace—as well as, more recently, as a digital ethnographic method for application online—but can be intrusive and problematic to apply to people's mundane and routine home life. In any event, the presence of a field observer in a private environment, such as a home, necessarily alters the environment to something less than private, with a concomitant effect on the social performances that take place within, and on the veracity of studies of connected homes. Moreover, and importantly, researchers are likely to want to hear the voices of the householders. Reflective, semistructured, and open-ended interviews are one way of cultivating this voice; interviews remain an important part of the toolbox of empirical research methods, and we've found that talking with people is by far the most important source of insight. However, the unique challenges presented by the home as a private space, and by the dynamism and complexity of the connected home, call for additional methodological approaches and processes.

The presence of a researcher recording the everyday activities of family life in the private sphere of the household, ideally for extensive periods of time, is often both impractical and invasive (Hine 2000; Mackay and Ivey 2004). Not surprisingly, then, household media ethnographies have tended to focus on particular technologies, including pioneering research such as work on television (e.g., Morley 1986; Spigel 1992), and computers (Lally 2002). Where the focus is not on particular technologies, household media research has largely used a synchronic methodology attending to the present situation, rather than a diachronic approach that is sensitive to historical change and effects. Yet, focusing on single technologies and "snapshot" approaches neglects the ethnographic importance of studying the interrelations of social and technical life in domestic settings (Nansen et al. 2009, 2011; Wilken et al. 2014). In another turn, media research has migrated to the internet, with a range of ethnographic approaches exploring virtual, online, and digitally mediated communication as a way to access and analyze the contexts and meanings of everyday media use (e.g. Hine 2000; Miller and Slater 2000; Rheingold 2000; Nardi 2010). In one sense, these approaches are a pragmatic response to the problem of access, with digital environments offering a wealth of archival, interpersonal, and experiential information. Yet, with some notable exceptions (see

Miller and Slater 2000), digital ethnographies have tended to erase the spatial, temporal, and cultural contexts in which media use takes place. In particular, the interrelation of the domestic space as an ordinary yet significant site of everyday media consumption, practice, and meaning has been marginalized through this decontextualization.

Our research sought to contextualize media use in domestic spaces by using both established and novel digital ethnographic techniques, which draw on the use of mobile, software, and visual technologies that have been used to conduct online ethnographies. By redeploying these technologies into domestic spaces and research methods, our objective was to understand how household digital media environments were changing and how everyday practices were shaped with and through new digital technologies, such as broadband internet. This research was situated within this changing household media ecology, a context that highlights the value of ethnographic enquiries for exploring the dense and detailed interrelations of multiple technologies and practices in domestic space (Shepherd et al. 2007; Wilken et al. 2011). Yet, in this "polymediated" environment, difficulties remained in accessing and capturing everyday media practices using traditional research methods. Homes are familial, intimate, and bounded spaces that prove difficult for researchers to explore in sustained ways. In response to these challenges, we describe below a research methodology we developed, based on digital ethnography approaches, that used mobile devices, digital ethnographic software, and creative data-collection activities in order to overcome the need for researchers to always be present in the field. The approach that we developed, and have refined over the course of a number of the research projects informing this book, addressed these difficulties through a staged process—using traditional ethnographic techniques but augmenting them with something more novel: the "domestic probe."

Domestic probes are an adaption of cultural probes, a method developed by Bill Gaver and colleagues and refined by others such as Andy Crabtree and colleagues, in response to the problems of user-centered design (Gaver, Dunne, and Pacenti 1999; Gaver and Martin 2000; Gaver 2001, 2002; Hemmings et al. 2002; Crabtree et al. 2003; Gaver 2004; Graham et al. 2007). Cultural probes are particularly suited to investigating people's everyday lives in settings that are difficult to reach using social science methods, such as questionnaires, interviews, focus groups, or participant observation. Cultural probes permit the collection of data from sites where researcher presence is problematic, allow research materials to be collected over longer periods in multiple locations, and enable participants to provide samples of their own world in their own way. While cultural probes may appear to have much in common with diary studies or experiential sampling, they are intended to provoke greater participant engagement. Aiming to be provocative, and sometimes whimsical, rather than strictly instructive, cultural probes attempt to disrupt familiarity with quotidian activities and prompt participants' reflections. Cultural probes are designed to encourage and empower subjects to collect, share, and interpret data in partnership with researchers. They also encourage participants to reflect on their own practices and to make those reflections available to researchers. Across our work, we adapted the cultural probe approach into what we called the domestic probe approach by merging it with other techniques in order to support a more ethnographic approach of recording the everyday activities of family life over an extended period of time, and to include participants as active collaborators in creating and interpreting their use of technology in the home.

In essence, the domestic probe comprised a box of equipment given to the household to use in order to record and interpret their use of domestic technologies. They were kits of provocations and recording devices. The intention in using them was that people were invited to tell their own story of their relationship with technology, and to have fun doing so.

The precise contents of each domestic probe pack depended on the makeup and preferences of the households being researched, and our use of them evolved and was refined over the course of our research. For instance, in the Connected Homes project (2004–2010) the initial probe pack contained (Figure 2.1) local, national, and global maps to trace origins and destinations of communications; color-coded stickers to record each device's user and frequency of use; digital and instamatic cameras to record snapshots

Figure 2.1. Image of probe pack used for Connected Homes project (2004–2010)

of the routine and the novel in domestic life; diaries for each household member; a scrapbook for photos and jottings; and additional stationery supplies such as colored pencils, marker pens, glue, tape, scissors, and so forth (see Arnold 2004).

In a later stage of the Connected Homes project, the pack was expanded to include the above as well as the following items:

The "random sampler" comprised a mobile phone with a built-in camera and an unusual ringtone. When the phone rang (triggered by the researchers), the householders took an immediate "snapshot" of their technology use at that time, or they responded to a "mission impossible" (see below).

The "frustro-meter" was a whimsical artifact in the form of a foam hammer or club, used by members of the household to mete out a thrashing to badly behaved technology. A pedometer embedded in the device recorded a rough count of the blows delivered over the research period.

The "document-box" was a kind of household creative toolbox for critical household economic accounting. The box contained highlighting pens and annotating pens, dubbed "censorship marker pens," that were used to annotate statements of transactions and accounts from phone companies, internet service providers, newsagents, and what have you, that, once annotated/censored, were then stored in the box for discussion with the researchers.

An "inspiration trap" was a one-time, twenty-second, non-editable voice recorder. Householders were asked to use the trap for recording a "eureka moment"—that is, occasions when they were inspired by an insight, or a conclusion.

A "special purpose" camera was of a disposable camera with a set of twenty-six photographic requests—for example, requests for a photo of the emotional center of the household, a photo of the television's "evil nature," and so on.

"Missions Impossible," or secret missions, were sent to a particular member of the household, through the mail or through an SMS message, at times and occasions determined by the researchers. A "mission impossible" involved asking the recipient to complete a task and record the experience using probe-pack devices. The task might be to insist on a change of routine (no TV tonight) or to shift or disable a specific technology.

The final item was a "Connected Homes" junk-mail catalog that was delivered to each participating household (Figure 2.2). It was compiled by the researchers and consisted of some off-the-shelf technologies (e.g., images cut out of a home electronics store brochure); some antique or historical technologies (e.g., pictures of sliced bread, a flush toilet, old phonographs, radios, etc.); some fictional, techno-utopian technologies; and some services or experiences as opposed to technology objects (e.g., twelve hours of uninterrupted sleep, one month's extra holiday, falling in love, painting a beautiful picture), and so on. Each householder was given a budget from which to select a limited number of items. Once selected, one of the following occurred:

Each household member filled out an order form; or
The household had to negotiate for one item, and record the negotiations in the scrapbook; or
There was a contest to win an item (to win, each household member had to complete the form expressing in fifty words or less why this item had been chosen over the others); or

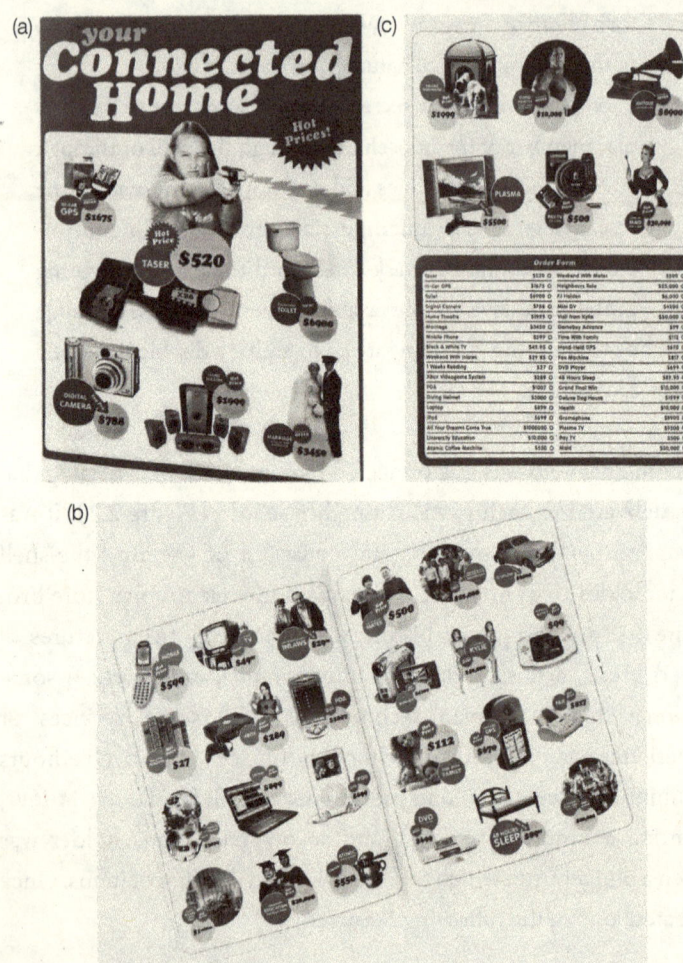

Figure 2.2. Image of Connected Homes junk-mail catalog (2004–2010)

The completed order form provided for multiple items, and the household (or each household member) had to fill in the form ranking the items in order of desirability (see Arnold 2004).

For subsequent projects, the contents of each domestic probe pack continued to be updated, all the while maintaining the underlying aim of providing a playful context and set of tools for creative or critical reflection on home technology adoption and use (see Wilken, Arnold, and Nansen 2011; Nansen et al. 2013a, 2013b).

By the time of the High-Speed Broadband project (2011–2017), we extended our participatory approach through the use of digital media. We deployed iPad minis preloaded with a data-collection software tool, EthnoCorder (Figure 2.3), which we had adapted for our domestic probes, extending the domestic probe approach with more digital ethnographic techniques. Participants used the EthnoCorder app, installed on these iPads, to record images, video, sound, and text and to store and share the recordings with the researchers.

Using these technologies, our participants were asked to collect situated visual representations of domestic technology use. In particular, we asked participants to periodically generate visual data framed around a number of playful televisual tasks. We describe these as "televisual tasks" because they were inspired by

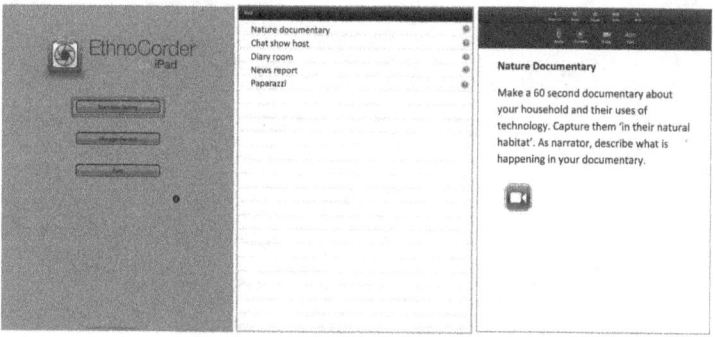

Figure 2.3 Image of Ethnocorder app interface (High Speed Broadband 2011–2017)

familiar television formats and conventions that would likely be familiar to participants. The tasks were designed to capture the household's technology use using familiar televisual genres of a "nature documentary" (make a short documentary about the household's use of technology and capture household members "in their natural habitat"), "news report" (create a short report about a piece of technology in the household [e.g., most loved, most frustrating, oldest, newest]), "paparazzi shot" (take a surprise photo of a member of the household while using technology and then, having shown the photo to him or her, decide on a caption together) (Figure 2.4), "diary room entry" (speak directly to the camera explaining what you think other household members' favorite pieces of technology were and why), and "talk show

Figure 2.4. Photograph of paparazzi shot, "couch office" (High-Speed Broadband 2011–2017)

interview" (interview another member of the household about his or her technology use, asking about habits, routines, and rituals) (see Nansen et al. 2016).

Participants were able to use the tablet camera functions to record still images and videos, and then the EthnoCorder app was used to store and upload the recordings for researchers to access. The tasks were designed to collect situated "glimpses of particular lives" (Boehner et al. 2007, 1082). They drew on familiar generic structures for producing, directing, and editing responses, at the same time offering households flexible and creative ways of describing their digital media use. The data they generated were intended not to be purely naturalistic but rather to provoke consideration and reflection about otherwise ordinary forms of media use. In an attempt to stay true to the sensibility of cultural probes and of our earlier domestic probes, the mobile devices, ethnographic software, and visual tasks aimed to collect specific fragments of data on household technology use and appropriation. We referred to these data produced by household members in subsequent interviews in order to explore subjective aims and interpretations of the recordings that were produced.

The domestic probe approach used physical or digital data resources that stood in for us as researchers *in absentia*. They remained behind after we left each household and provided these households with "objects to think with" (Papert 1980), conversation pieces, and grist for the mill of conversation between participants and ourselves. This is to say that the traces that we collected were not "evidence" as such; rather, they served as conversation starters and stimulants for reflection (see Arnold, Gibbs, and Shepherd 2006). In this way, the packs fulfilled an even more important function: By agreeing to participate in the study, the householders were in effect agreeing to participate as *co-researchers* or *collaborators* in our research work. Although as researchers we recorded, interpreted, and analyzed the probe traces and the ensuing conversations, the probe required the close collaboration of the participants, not just as passive data sources—as subjects of research—but as full participants. Not only were they responsible for the traces that built up as the probe was used, but the probe's traces invited participants to reflect on and articulate their relations with the technology as the traces accumulated. The focus of analysis was not

> the material artefacts of the probes—the tapes, the photos, the booklets and diaries, etc.—but rather, the *situated character* of everyday life ... elaborated by *participants' accounts* of their daily rhythms, routines, and abiding concerns. Such accounts supplement and augment insights gained from direct observation and are generated through cooperative analysis of the returned probe material. Probe materials serve as triggers for analysis then and in asking people to administer them we transform participants into active enquirers into their everyday lives, rather than passive subjects of our research. (Crabtree et al. 2003, 7, emphasis in original)

In most cases, we met with each participating household three times. On the first visit we dropped off the domestic probe pack, explained its contents, and conversed generally about the household's technology use. Some two weeks later we returned to be taken on a "technology tour" of the home, to be led from one room to the next by one or more members of the household, during which we filmed the domestic media ecology in situ as the participants told stories about their origins, their history, their use, and their strengths and shortcomings. Similar to "object biographies" (Silverstone and

Hirsch 1992), the technology tours involved householders walking the researchers around their homes and identifying what technologies they own, and describing the ways and reasons they are used by members of the households. The method was useful for eliciting audits and accounts of technologies prompted by their presence rather than relying on participants' recall. This approach supported conversations between researchers and participants where the latter explained the significance of broadband for their households (the primary objective of the High-Speed Broadband project 2011–2017) and facilitated understanding the ways in which existing and newly introduced technologies were being used in private spheres of everyday life. Our third and final visit took place in earlier projects some three to four weeks later. However, in the High-Speed Broadband project (2011–2017), we visited participating households yearly, for three years, and deployed the domestic probes three or four times between visits. During subsequent visits we engaged in conversation relating to the traces captured by the probes.

In effect, the recordings or traces generated through the use of the domestic probes provided provocative and evocative grist for the mill of conversation among the participating householders, and between the participating householders and ourselves. The conversations and the interpretation of probes circulated around the subject–object relations detailed earlier. We talked about the media and technologies that populate the house, why those and not others, why "there" and not "here," what habits and rules of use pertain, how they are brought under control, who does this, what attitudes evolve, what roles are played out, what emotions arise, what purposes are achieved, what human relations are affected, what values are relevant, and what behavior results.

CONCLUSION

The insights yielded by these domestic probes, when refracted through the theoretical prism of media and communication ecology, reveal a range of socio-technical and socio-material complexities of lived relations and dynamics pertaining to technology use within the home. These complexities coalesce around a series of pressure points or key themes that form the focal points of investigation of the chapters that follow. As noted in the previous chapter, dense ecologies of media are creating environments where media screens, connection, and communication are persistent and ambient within contemporary homes. And yet these household media ecologies are constantly evolving and subject to ongoing processes of settlement and resettlement.

The remaining chapters of this book examine a range of issues that speak to the challenges households face in negotiating technological change and flux within media-rich homes. Chapter 3 explores how the media ecology of the home is socially and materially reshaped over time to accommodate multiple technologies and, in particular, how the reconfiguration of spatial practices and meanings emerges through the steady accumulation and distribution of digital media over the course of this century. Chapter 4 examines the extensive "articulation work" needed to bring coherence to and management of the domestic media ecology, and the efforts involved in maintaining the proliferating array of digital technologies within contemporary households. Chapter 5 explores the negotiations that take place within the household media ecology around appropriate technology use, with emphasis on the complicated rhythms and temporal dynamics at play in the domestic media ecology. In Chapter 6 we suggest that digital domesticity can also be shaped in significant ways by varied modes of *not using* household technologies. Our argument here is that, as with use, non-use cultivates household media ecologies in myriad ways that emphasize the relational, material, and interdependent nature of the contemporary, digital, networked home. The final chapter of the book explores how and when older things—such as technologies that deteriorate, break, or become obsolete or no longer valued— are repositioned within and beyond the household media ecology.

In the following chapter, we analyze how various home spaces and their associated practices have been remediated, amended, and reconfigured by the changing domestic media and communications ecology over the course of this century. We discuss how practices of dwelling with media, which revolve around accumulation, diffusion, and mobility, intersect with the spatiality of the home. In exploring these issues, we focus on some key spaces, including the kitchen, the bedroom, the living room, and the bathroom. Of particular interest to us in the next chapter is the relationship between the human inhabitants and the media occupying the house, how inhabitants rationalize and reconcile the accommodation of new media into existing media ecologies, how their media use adapts to the physical spaces and practices of dwelling with media, and how they adapt to the changed ecology resulting from a changing ecology. This suggests a concern for materiality reminiscent of what Elaine Lally (2002, 25) describes as cultural proprioception, in which the placement and use of technologies provide a kind of spatial scaffolding that supports and structures our sense of place in the social world.

[3]
APPROPRIATIONS

This chapter considers how the landscape of home media technologies has shifted over the early stages of this century, showing how relations between devices and householders create dynamic and evolving media ecologies. The chapter describes a raft of media technologies and internet infrastructures that have been developed for the home over the last fifteen years and examines the motivations and strategies for bringing these technologies into the home. It explores how new media are incorporated and appropriated into domestic spaces and everyday practices and, in turn, how particular domestic spaces and practices are adjusted and adapted to these new technologies. It considers the rationales for appropriation on a scale that embraces global cultural and economic forces through to the particularities of locating and dwelling with media. These technologies are not just located in the home; rather, in important ways they constitute a place as a home. They form the sociotechnical environment in which we live, and their particular configurations have implications for what goes on at home, for our day-to-day lived experience.

The chapter is necessarily coarse-grained. The growing number and constantly changing nature of technologies, along with differences in household demographic factors, geographic

Digital Domesticity. Jenny Kennedy, Michael Arnold, Martin Gibbs, Bjorn Nansen, and Rowan Wilken, Oxford University Press (2020). © Oxford University Press.
DOI: 10.1093/oso/9780190905781.001.0001

locations, income, lifestyle habits, and everyday behaviors, are all too various to track in fine-grained detail. Instead, we aim to provide a landscape view that maps the broad trends that have shaped the domestic media and communications environment in all its variation, through the living room, the home office, the media room, bedrooms and other "no-go" zones, and the smart home, finishing with the appropriation of the key precursor technology—broadband. Informing the description of this changing landscape are the notions of the house, the home, and the act of dwelling, and informing the changes that have occurred in this landscape are the dynamics of diffusion, accommodation, and appropriation.

DENIZENS OF A CHANGING ECOLOGY

Over the past fifteen years, households in the developed world have undergone a drastic transformation in the population, diversity, and intensity of domestic digital media and communications technologies (Wilken et al. 2014; see also Chapter 1). At the turn of the present century, a typical domestic media ecology comprised a cathode-ray television in the living room displaying a limited number of free-to-air channels, and perhaps connected to a videocassette recorder; a desktop computer in a home office, perhaps connected to a dial-up modem; and a landline telephone, in most cases located in a communal area in the home. In more recent years the home has become a place for high-definition smart televisions, multifunction set-top boxes, high-speed broadband, wireless home networks, mobile computing, cloud services, online commercial and government transactions, gesture-based interfaces, personal and portable entertainment systems, streaming media of all kinds, and, most recently, voice recognition, virtual assistants, and smart home devices.

Our ecological approach suggests we should consider the ways in which people, domestic artifacts, household architectures, and new technologies are materially, spatially, and temporally woven together to constitute the particular kind of place called home. Firstly, we may consider the domestic space as a *house*—as physical infrastructure that is made of bricks or wood, that is material and obdurate, and that is distinct from the people and performances that inhabit and enliven it. For other purposes we may consider the domestic space as a *home*—a place that is lived in; the physical space that is at one and the same time a social space where people and the artifacts that define a place as a home dwell, and is understood in our culture to be an important locus of social interaction (Saunders and Williams 1988, 82). Then again, a useful unit of analysis may refer to the people, to the *householders* who dwell in a house, thereby constituting it as a home through interaction with each other and through interaction with non-human actors, important among which are media. And fourth, the performances of the house, the home, and the household may be thought of as coming together in the act of *dwelling*: the practice of living (being) in the home. Each of these ways of imagining domestic space privileges a particular cast of actors—the *house* focusing on a material construction, the *home* on the interrelations between the social and the material, and *dwelling* bringing the whole cast together—house, home, and household.

While the study of this ecology is often centered on the *human* occupants, it is increasingly difficult to ignore the role of the *non-human* occupants—the ever-increasing volume of media *stuff*, each of which has demands of its own. For example, the historical work of Witold Rybczynski (1986) showed how industrial technologies of gas, electricity, and telephones strained and changed the structure of houses by demanding their own space and affecting the architecture and uses of the home space. The media history work of Lynn Spigel (1992) has traced how the structure of the American home was further redesigned to "make room" for TV, and the shifting practices and social ideals of dwelling it proposed. Elizabeth Shove's cultural geography approach has pursued these ideas through her materially oriented approach to the co-evolution of standards for comfort, cleanliness, and convenience in the home (2003)—mediated through technological innovation (e.g., freezers, washing machines) and through studies of home extensions (Hand, Shove, and Southerton 2007). One of Martin Hand, Elizabeth Shove, and Dale Southerton's key findings was that common explanations for home extensions revolved around the need for more space, often to accommodate the accumulation of stuff, and the modes of dwelling that new technologies implied for "accomplishing what people take to be normal, ordinary, and acceptable ways of life" (Hand, Shove, and Southerton 2007, 678). Thus, things do not simply take up room, but impose new demands on domestic practices:

> The diffusion and ownership of increasingly standardised appliances have the unintended and unexplored effect of scripting and thereby standardising the routines and habits of those who use them ... people extend their homes in order to accommodate additional appliances and more importantly in order to accommodate practices inscribed in technologies. (Hand, Shove, and Southerton 2007, 679)

For Hand, Shove, and Southerton, then, it is not simply the accumulation and turnover of possessions that change the *house* that is important for studying domestic material life, but rather the way that technology changes within the house change ways of *dwelling* in the *home*. They argue that "domestic technologies are implicated in the structure and reproduction of practice and hence in the choreography of things and people in time and space" (680). And, in their study, innovations demand changes in dwelling in the home (particularly in kitchens and bathrooms) through processes of *multifunctionality* (in kitchens) and *multiplication* (of bathrooms).

When it comes to new forms of digital media, similar patterns of multifunctionality and multiplication are evident, but, in this case, changes apply to the entire house, not just individual rooms such as the kitchen or bathroom. So, an understanding of the domestic media ecology is captured not by their impact on the multifunctionality and multiplication of defined spaces alone, but also by the *distribution* and *mobility* of practices these technologies imply. In the case of media technologies that are distributed through the home and are mobile within the home, we see forces at work that are different to those that apply to freezers, en-suite bathrooms, and microwave ovens. The imperatives to purchase new things that are motivated by commercial obsolescence and innovation (Parikka 2011b, 2014) apply to media technologies perhaps more so than to home appliances, but these imperatives are now joined by demands that are peculiar to media and communications technologies: the demands of shifting employment and working practices that require constant connection (Gregg 2013) or of shared cultural norms, tastes, and expectations about constant connection in everyday life.

The logic of consumer capitalism is particularly evident in the casual or unreflective acceptance and accommodation of waves of fast turnovers of stuff like flat-screen televisions, sound systems, wireless routers, desktop and laptop computers, games consoles, mobile phones, and the like. A more instrumental logic of rational problem solving is evident in views of media as offering solutions to perceived problems or needs. Here, the solution is legitimated by reference to the objectification of some quality or value (say, domestic ergonomics, return on investment, energy reduction, enhanced security, and so on). Finally, however, such human-centered logics are often undermined by the socio-material entanglements of accommodating and dwelling within dense media ecologies, in which complexities and decisions unfold through more distributed and interrelated arrangements—or, in the words of Michel Foucault, *dispositifs* (1980). These three lenses are described and examined below, drawing on empirical evidence gathered over the last seventeen years and discussed in the opening chapter, and organized thematically in relation to three key terms: *diffusion, accommodation,* and *appropriation.*

Diffusion

Theories of diffusion (e.g., Rogers [1962] 2003) have been influential in commercial and popular analyses of new technologies, including domestic media, in terms of identifying and characterizing widespread patterns and cycles of adoption. The model describes the diffusion of a given technology through a population through time in terms of a spectrum ranging from early adopters and innovators (who are the first to take up a new technology), to early and late majorities (who come later and are the mainstream market innovators look for), and, finally, to laggards (who take up the technology last, if at all). This widely accepted model claims to capture a predictable and orderly pattern of technology adoption, which moves in a clear direction from invention or innovation to widespread adoption. The diffusion model suggests a certain inevitability, a kind of technological determinism that begins with the assumption that a new technology is in some clear way superior to the old technologies, and, if so, moves in a linear fashion through early adoption to widespread adoption with the assistance of appropriate marketing. Aligned with the logic of consumer capitalism, the diffusion model was developed and applied in the post-WWII era of mass production, mass marketing, and mass consumption. And indeed, the new automobiles, electric refrigerators, and cathode-ray televisions of the 1950s did diffuse through mass markets in roughly the way the model suggested. In more recent times, though, the limitations of the model have become evident, and its application to waves of fast-turnover media is limited. Today there is no single mass market through which products might diffuse, but a series of highly differentiated markets, with each making different demands of products. There are no single products produced and marketed on a mass scale to the population as a whole, but a myriad of finely differentiated product-models, each customized to appeal to these niche market segments. Moreover, the model assumes that any given technology has a static ontology—that a technology is what it is, and what it is and what it does are fixed at the time of design and manufacture and do not change in the context of use and the duration of its life cycle, not to mention its place within a wider complex of competing and alternative technological products. Today's market segments and niche product specifications change very rapidly, a phenomenon not captured by a model in which take-up through time changes but products and markets do not. In these ways, the model elides the ways new technologies become embedded in today's technical, social, and economic contexts, including the media ecology of the home.

Accommodation and Appropriation

At the beginning of the twentieth century, the home was rearranged as a place that was made for telephones, radio, and gramophones. In the middle of the century, it changed again for television; in the 1980s, for home computing; in the 1990s, for the internet; in the noughties, for wireless streaming; and, most recently, for the proliferation of mobile screens. Broadcasting, first via radio technology then television, created and met a new social need to bring professionally produced news and entertainment into the home (Williams 1975), thus providing families with a window on the world from the living room (Spigel 1992, 7). Decisions had to be made about where this new technology belonged. The arrangement of seats was now directed toward the television instead of toward each other, toward a fire, or around a radio. Early televisions were not just a screen—they were furniture and were manufactured within timber frames, with often ornate craftsmanship. In these ways and more, the demands of the television for space, for position, and for attention were accommodated. The house was changed for the television; its use provided an important social experience as well as a solitary experience, thus changing the sociality of the home (Morley 1986); and the social and material accommodation of the television changed the characteristics of dwelling. In this new way of dwelling, other objects in the living room, such as pianos, radios, and radiograms, had to make room for television by being replaced or being displaced to more peripheral spaces. Competency requirements for television use were low, unlike the piano or the earliest personal computers, and constituted no barrier to accommodation. New packaged foods formed an alliance with the television—"TV dinners" that were quick to prepare, came in disposable foil, and could be consumed while watching television from an easy chair—and this had significant implications for the dining room and the kitchen. This new material configuration of the living room and its spillover effects on the functions of the kitchen and dining room changed the house, materialized a different home through resourcing different forms of socializing, and thus engendered a new form of material-social dwelling.

But this is not to say that television reshaped each house, home, and dwelling in the same way. The demands of television were not simply accommodated, but rather the television was appropriated, which is to say that each house and home found its own way of dwelling with the television. To put it simply, the television makes demands that need to be accommodated (for space in a room, for an orientation to furniture in the room, for electricity, for attention, and so on), but the household also makes demands of the television, and in this way the television is appropriated. In some homes and in some circumstances, the television was always on and provided a constant backdrop of sound and image. In other houses in other circumstances, the television was allowed to participate in home life only at set times. In still other homes, the television was used by family members differently; for example, children and adults dwelt with the television in different ways. And, so, while the television was accommodated in all homes that had a television, it was accommodated in quite different ways—spatially (in the house) and functionally (in the home) according to the particular characteristics of the ecology the television was joining—a concept signified by the word *appropriation*.

These more situated notions of accommodation and appropriation were initially captured in theories of domestication (e.g., Silverstone and Hirsch 1992; Silverstone and Haddon 1996), discussed in the previous chapter, and their empirical application to analysis of media adoption in the wake of the diffusion of television in the second half of the twentieth century. The vocabulary and orientation of the domestication literature aimed to consider both the practical and the symbolic aspects of the adoption and use of technologies, showing how these two elements, the meanings of things and their materiality, are intertwined and become part of everyday domestic life in which environments and users change and adapt. Studies of "media homes," informed by domestication approaches, were established over the course of the late twentieth century, to discuss the ways different waves of media technologies, such as televisions and computers, became physically and symbolically located within home spaces (Silverstone and Hirsch 1992; Silverstone and Haddon 1996). This approach has, then, been helpful in complementing diffusion approaches centered on broad social trends of technology adoption by examining the contextual arrangements for appropriating and accommodating new technologies into the home. Yet this approach has remained limited in its focus on singular and particular technologies rather than interrelations of multiple technologies configuring the home media environment. And, while it turns somewhat to the materiality of media, this attention is overshadowed by a focus on the social and symbolic dimensions of adoption. While the domestication research has predominantly focused on particular technologies, our research builds on a socio-technical tradition of studying more recent domestic media technologies, highlighting the importance of studying the communications ecology that increasingly involves dense interrelations of multiple technologies configuring the home media environment (Shepherd et al. 2007; Nansen et al. 2009, 2011; Wilken et al. 2014).

We suggest these limitations in scope can be further developed by looking at how domestic spaces and practices are adjusted and adapted to the aggregation of new technologies and infrastructures—that is, by considering innovations and their adoption in the contexts of the aggregated home technology environment. In this book, we examine how since the turn of this century we have seen Australian households, as elsewhere around the world, undergo a drastic transformation in the number, variety, and intensity of digital media and communications technologies, and an associated transformation in the way people use and interact with media technologies in the home. Below, we discuss how this ecology has changed in response to the adoption and use of multiple media infrastructures, technological artifacts, and digital media products in the home, analyzing how various home spaces and their associated practices have been remediated, amended, and reconfigured by the changing domestic media and communications ecology over the course of this century.

THE LIVING ROOM

This century, the house, the home, and our way of dwelling have been shaped by the appropriation of a new wave of media technologies no less powerful than last century's television. Over the past ten years, wireless connectivity and mobile screens have colonized the home's ecology, resulting in a decline in specialized, dedicated, media-specific rooms and spaces. Within contemporary broadband- and wireless-connected household internet infrastructures, locating particular media in particular spaces is far less common than in the past, changing the way the household dwells with media. Rather than finding a niche in the corner of a living room, or in a study or office, media devices are now more likely to occupy any and every household space. Rather than being appropriated by groups of householders, media devices are appropriated by individuals. In these ways, today's domestic media are ubiquitous. To dwell with this media is to be entangled in a media ecology that is insensitive to time and space, in that it is potentially everywhere and anywhere, anytime, and is less sensitive to domestic sociality insomuch as it is accessed and experienced by individuals and not household groups.

At the turn of the century, television viewing was largely contained within specific and limited domestic spaces. In research we conducted prior to the introduction of broadband, with slower internet connections, such DSL and ADSL, along with fewer homes using wireless routers and mobile devices, we saw a stricter regime of spatial organization and management around household media. Mary and John, for example, took this approach by deliberately not having a television, telephone, or computer in the open-plan living room and kitchen, which was the space where the family spent most of their time. The only TV was in a separate room—a space dedicated for television viewing, and their viewing was planned and purposeful: "Our view is that we only want TV in here, and not out there where we are interacting and having dinner" (Mary, Connected Homes 2004–2010).

Then, we saw broadband infrastructure and mobile media ecologies contribute to more expected and now familiar forms of media use develop, such as more customized TV and entertainment viewing, creating more personalized rather than corporate experiences of flow (Williams 1975) through the combination of faster internet connections, streaming services, digital video recorders, and so on:

> The main thing we notice it in would be YouTube and streaming stuff, there's no buffering. The other thing is when we download TV shows and stuff. We used to always download them in advance and let them go overnight, but now if something new comes out we just download it and then 2 minutes later we can watch it. (Joel, High Speed Broadband 2011–2017)

The families we observed were slowly updating their televisions, as newer digital flat-screen televisions entered the market, making older analog cathode-ray sets feel obsolete with their cumbersome size. Upgraded digital flat-screen televisions were usually placed in the same position as the obsolete analog one. Older TVs (until they became dysfunctional with the switching off of the analog network) often migrated to other places in the house, often children's bedrooms. While in some instances this initiated processes of distributed viewing, and critiques of "living together separately," we also observed unexpected contradictions, with the screen "real estate" of the larger flat screen, along with the emergence of higher-definition picture quality, drawing family members back to shared viewing. When asked how the new television had changed the use of this room, John, one of our participant parents, noted that "we find that the kids come in a lot more" (Connected Homes 2004–2010). Despite having a personal television in his room, the superior quality of the flat-screen TV often enticed the children back to watch television with their parents, sometimes reinstalling older traditions of collective family viewing.

But these ideals were not uniformly realized and were contradicted by the new and novel ways viewing devices, spaces, and times were redistributed. Key during this time was the emergence of online video-sharing platforms like YouTube (Burgess and Green 2009), along with the evolving functionality of devices, with computers becoming increasingly used for entertainment as much as work activities. During our earlier research, audiovisual entertainment was consumed solely on television and was necessarily confined to the television room (Wired Homes 2002–2006); now, laptops were also appropriated for audiovisual entertainment, no longer located uniquely with the television, but distributed and mobile. Viewing devices had multiplied. Digital content could be consumed on different devices, in different places, and at different times, previously set aside for activities such as eating or chatting, breaking down spatial, temporal, *and* socio-technical distinctions. This had an impact on dedicated spaces for relaxation, entertainment, and television viewing, which in many ways remained spaces for viewing television but were no longer spaces for sharing this activity. Instead, multiple media were used simultaneously, yet separately, within the same spaces. As John noted, "Now with the laptops, there is now a lot of internet use, so that is a different use of that space . . . Mary might have her laptop and Angela might have her laptop . . . not really sharing, except bandwidth" (Connected Homes 2004–2010). But the blurring of activities within spaces meant that while the living room became a site for more customized or even individualized viewing of entertainment content, this space could also be used for work rather than having to decamp to the home office:

> The couch becomes an office space when sick. The funny thing about work nowadays is that you sit down sick and you log on and chat and do a little bit of work, unless you are on your deathbed. Or if I am just at home on the kitchen table. (Nysha, High-Speed Broadband 2004–2010)

While multitasking may not necessarily be new, certainly the increasing speed of the internet, alongside the mobility of devices, is contributing to a situation that affords new configurations, patterns, and intensities of multiple screen activities. The entanglement of ubiquitous media with modes of dwelling is a messier business (Law 2004) than is suggested by the diffusion model, though concepts of domestication, accommodation, and appropriation do somewhat better. The situation is messier because few spaces are now the exclusive territory of certain classes of media, and few times are set aside exclusively for the use of a particular device or activity. However, though wireless and mobile technologies are dispersed throughout the house, and single-purpose spaces are less common, in most of our cases the living room remained the key center for the collocation of broadband media, and the couch and the main screen were the poles around which other technologies oriented.

In the High Speed Broadband project (2011–2017) homes, anchored by the couch facing the screen and the screen facing the couch, we found a miscellany of co-located laptops, pads, game consoles and controllers, smartphones, set-top boxes, routers, and speakers. A proliferation of screens and media technologies occupied the living room, making it a place where the co-location of media translates into co-location of activities and individual householders browsing the internet, checking social media feeds, watching YouTube, doing some work, listening to music, or making online purchases, each according to their own desires:

> Well, it's been good because, like I said, I'm not having to go upstairs to the computer, because that's where we've got the computer. Just, got to Google something up quickly or even just do banking. Scott will use eBay, he gets onto eBay, he's using his phone a lot. Of course, on his phone you can't see the things properly because the phones are a good size but they're not big enough. With the iPad, it's that in between [size] where it's just a good enough size where you don't have to go upstairs on the computer to see everything. Just limits how often you have to sit upstairs. You feel a little isolated when you're sitting in the study. At least if I'm sitting down here, I might be watching a movie, he might be looking up something, but we're sitting together.

Almost like having a laptop without feeling like you're actually doing work. (Diane, High Speed Broadband 2011–2017)

Why is it that a miscellany of screens, PCs, keyboards, game consoles, controllers, DVD players, routers, couches, and speakers congregated together in the living room? Established diffusion approaches might say "because the household decided to adopt these technologies and decided to put them there!" and then set out to provide an explanation for this in terms of human desires for a certain mode of dwelling, analyzed in terms of efficiency and functionality, cultural aesthetics, the semiotics of mess and order, gender politics in the home, psychological motivations, and the like. Such an analysis provides only a partial account insomuch as it elides the material requirements of the technology itself; that is to say, the technologies have requirements that need to be *accommodated* before the technologies can be *appropriated* for household use. For example, an account that recognizes the material agency of the technology demands that they be co-located as a precondition to meeting human requests. The nearer the laptop is to the wireless router the happier the laptop is with its position, and if the laptop is not happy with its position in relation to the router, the human user is not happy in its relation with the laptop:

Certainly, the NBN [household internet] is faster, there is no doubt about that . . . but the Wi-Fi diminishes the speed of the service undeniably. If I bring my notebook down here and put it right next to the modem it's a lot faster than if I work up in my office. (Graham, High-Speed Broadband 2011–2017)

Most game consoles, controllers, screens, and the like need to be co-located in close proximity in order to interact, and their interaction among themselves is a prerequisite for our interaction with them. Although objects do not move themselves to the living room, they do tend to congregate in the living room in order to be with us and in order to be with each other.

THE HOME OFFICE

Rooms in homes dedicated to work have a historical role in the appropriation of the bulky, wired, and fixed PCs and peripherals of the late twentieth century. The "home office" was a relatively recent material inclusion in domestic architectures and was also a nascent lexical expression, replacing the "home study" as a space dedicated to work within the home (see Shepherd et al. 2007). This space was typically marked by a density of work-related technologies that reflected the growth of distributed, remote, and nontraditional forms of labor, and their inclusion within domestic environments. The creation of a dedicated workspace within the home was an example of rigidly ordering, segregating, marking, and defending work space from family space, and vice versa. Tom, for example, maintained a distinct working space through material boundaries. His office was a designated space, which signified work as well as enabled work. He only worked in the home from this designated space despite having wireless broadband throughout the home, and despite owning multiple laptop computers because he thought his productivity "would go down too much" (Tom, Connected Homes 2004–2010). Others also made clear attempts to spatially separate paid work from family life within their home through the thoughtful location of their technologies. Mary and John's study, for example, had three desktop computers, two of which were used regularly; all were networked and connected to a scanner and printer (Connected Homes 2004–2010).

Yet, as ostensibly dedicated spaces for work, it was also common to see these kinds of rooms crowded with technologies, both functioning and obsolete, both digital and analog, including radios, landline telephones, old televisions, books, and magazines. They were, then, in some sense, also designated spaces of storage for devices and artifacts loosely affiliated with technology, or unsure where to be housed elsewhere in the home. Thus, while nominally dedicated working spaces accommodating work-related technologies, they were also graveyards for media technologies.

As we have seen, though, over the course of this century dedicated spaces for particular media and particular activities such as work have become increasingly uncertain and unsettled:

> No one says I have to put in the time at home—it's more implicit than explicit—but they know that I'm not going to be able to meet particular deadlines unless I do it . . . The fact that we have cable internet, plus the appropriate desktop-publishing software, means that the boundaries between work and home become fuzzy. But it also enables me to be excellent at my job. (Mary, Connected Homes 2004–2010)

While the architecture of the house in the early part of this century required a home office or a study to accommodate the bulky and fixed-in-place desktop computer, modem, printer, filing cabinet, work desk, and the like, this historically important role seems to be a thing of the past. In contemporary household media ecologies, such spaces still existed but were less inviting than the living room for media use:

I would hate to be in front of the computer of a nighttime. I cannot for the life of me imagine anything more awful. [Jørgen's] study is a completely uninviting room, like it's a not nice place to be. It's cold, it's out there, it faces a wall. We've never done anything about making it comfortable . . . And we don't want to be there. (June, Connected Homes 2004–2010)

The home office, then, gave way to the living room (and the kitchen) as the place to interface with the wider world through media—which is not to say that home offices disappeared. Rather, home offices, studies, or guestrooms were repurposed as adaptive, flexible object-spaces, frequently accommodating spare furniture and miscellaneous stuff, along with the remnants of its important history—a desktop computer, modem, and printer that hadn't been used in years, along with shoeboxes full of cassette tapes and CDs, all sitting on a dusty work desk. Indeed, while messy, the order that existed in relation to the couch and the main screen in the living room was relatively exemplary in comparison to the lack of order that frequently existed in the home office. Figure 3.1 shows a stripped-down and stylized diagram to document the disorder typically associated with the PC, the work desk, and related objects.

The dedicated functionality of the home office was not so much a consequence of human requirements for a way of dwelling but a consequence of the need to accommodate the requirements of the desktop computer and its companions, requirements that have now been subverted by the provision of services that facilitate the distribution of work throughout the house. And, yet, these rooms gained a distinct new character as the ecology's graveyard for redundant technologies, and as sites of storage and clutter.

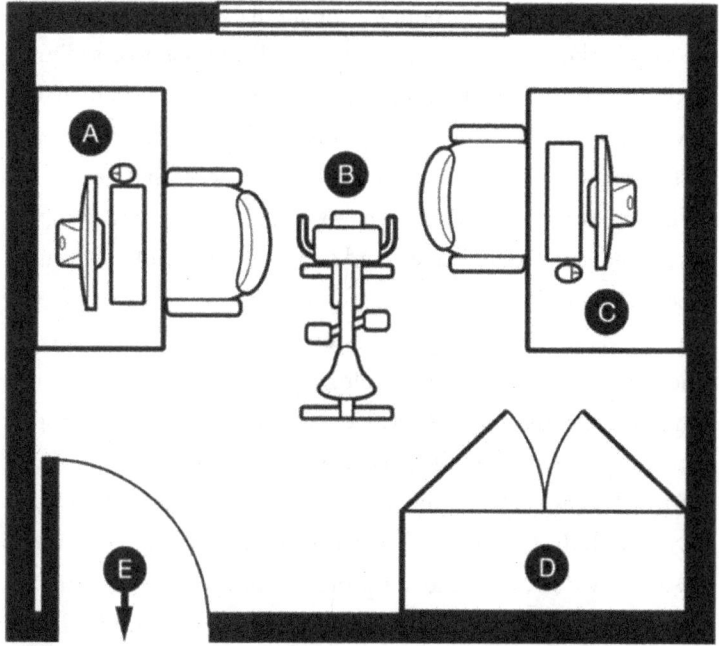

Figure 3.1. The "messy" study of Jeremy and Amy's home (High-Speed Broadband 2011–2017). (A) Main desk and PC. Has large stacks of paper on it, but otherwise clear of objects that prevent use. (B) Large items in the middle of the room, including an exercise bike with clothing hanging off it and boxes. The items need to be shoved out of the way to sit at the second desk (C), which has boxes and miscellaneous items on it, including a potted plant. (D) Cupboard with server, modem, monitor, and network devices. The equipment for broadband connectivity and battery pack is located above this cupboard. (E) Door to hallway. *From field notes, High-Speed Broadband 2011–2017.*

THE MEDIA ROOM

The "media room" of the early years of this century was an echo of the strategy that householders used to discipline unruly work-related technologies by confining them to the home office or study. Some media rooms were very elaborate "home cinemas" equipped with a large-format flat screen or high-definition projector, surround sound, and reclining chairs oriented to the screen. Others were less elaborate. For example, one family we visited lived in an old weatherboard house that displayed all the material sedimentation that attests to dwelling (Connected Homes 2004–2010). It was a lived house, not a show house defined by clean lines and neat surfaces. The media room in this house lacked state-of-the-art technologies but warranted the name "media room" because it was a place dedicated to the screen, and to music. It was not a place used for entertaining visitors or doing homework, and the seating was configured for viewing rather than for conversation. There were no other televisions or sound technologies in the house. That is, in this home the television and the video were not ambient background features of a multipurpose space, ready to be called on at a whim, or turned on as a default position to serve as background for other activities. Rather, media were appropriated in conscious decisions to watch a video or a broadcast program, and parents and children would relocate purposefully and mindfully in order to do that. It was a floor plan for purposeful, deliberate action that brought together media and parents and children in a controlled way.

During our first visits the householders talked of the upcoming renovation to the structure of the house in which priority was given to the media room in the new floor plan. When we returned three years later, though, following the renovation, we found that the media room had disappeared altogether, and all the media were located in an open-plan living space (Figure 3.2). A house structure that enabled integration rather than one that enabled differentiation was now the preferred mode of dwelling. Subsequently, media consumption, paid work, entertainment, and home life were more integrated—overlapping within the same floor plan. However, this change of logic did not signal a strategic change to idealized modes of dwelling, and like the old floor plan, the new one was consciously undertaken with parents, children, and media in mind:

APPROPRIATIONS

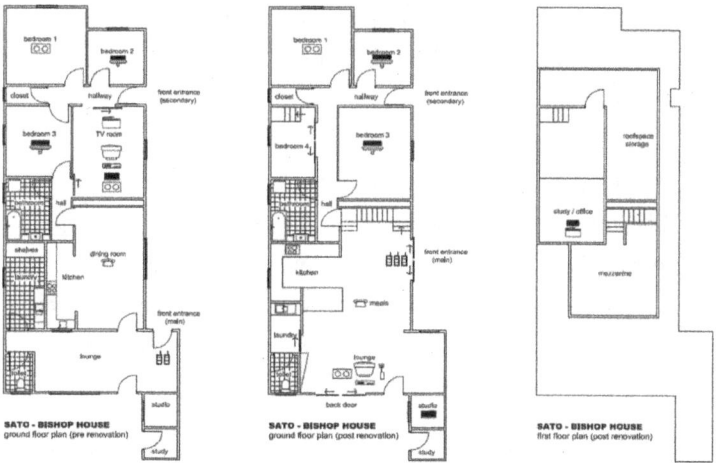

Figure 3.2. Floor plans before and after the renovation; the media room was removed as part of the renovation (Connected Homes 2004–2010)

> We wanted to have plenty of spaces where we could put a laptop and we could have the television. We don't particularly want to have a TV in another room because otherwise the kids would be in there and we would sort of lose them. (Ruth, Connected Homes 2004–2010)

The shift to more mobile media devices at this time meant that dwelling with media became increasingly uncertain and unsettled. Where and when activities such as communication, entertainment, or work were undertaken and the functions served by various devices had implications for dwelling. Spaces were less able to be set aside for specific or segregated activities, and families were less willing to do so. Instead, as one family answered, their dwelling with mobile media was "anywhere and everywhere" (Mary, Connected Homes 2004–2010). Importantly, such ecologies were also underpinned by the growing prevalence of houses that were wirelessly networked, and so the use of computers (for work, entertainment, and organization) was made mobile and distributed throughout the house. John (the father) said:

> Wireless network[ing] in the home has made it completely different how we operate. It is much easier to have multiple computers, and is the fact that we now have got multiple computers... it's just amazing how you become accustomed to a technology, and accustomed to doing things in certain ways. (Connected Homes 2004–2010)

BEDROOMS AND OTHER NO-GO ZONES

Thus, many contemporary media technologies found their way to the living room, and many redundant technologies found their way to the home office, but other spaces were somewhat hostile to technology and in some households were "technology-free zones." Both adults' and children's bedrooms were commonly spaces where computers, televisions, and to a lesser extent phones were excluded. Earlier in the century it was not uncommon for families to exclude phones—both landline and mobile—from their bedrooms:

> I turn the phone off if I want to rest. I pull out the plug on the other phone [landline]... I pull it out at night so that no one rings me during the night... people will ring you back... our philosophy is, "what is that critical?" (Jacka, Connected Homes 2004–2010)

Steps to manage and limit media in certain spaces were related to maintaining the distinct qualities and meanings associated with particular spaces (like the bedroom or dining room) or times (like the weekend, in addition to technology-free times, when/where communications technologies are turned off). Therefore, we also saw, during this period in the earlier part of this century, technologies consciously placed to help separate different areas and activities. Carving out technology-free times was a popular strategy in households at this time, over both short and long periodicities, which extended beyond turning technologies off to ignoring technologies if they beeped or rang, particularly during significant or valued family events, such as weekends or mealtimes. Thus, the practice of "switching off" enabled participants to resist the imperative of media and to carve out moments of pause or refrain against an increasingly mediated background rhythm, especially around important and valued spaces within the home. In John's bedroom, the only piece of technology was a clock radio, which, he noted, "doesn't do much but wake us up in the morning" (Connected Homes 2004–2010). The only landline phone in John's home was located in the hallway.

But, over time, the mobility of contemporary media eroded this embargo, and many felt the pull of always available news and information streams, checking social media feeds just before going to sleep or first thing on waking. As Dawn said, "The iPad lives within arm's reach, the computer usually on floor next to bed. [The iPad's] handy, I can write reminders to self if I remember something during the night" (High-Speed Broadband 2011–2017). Media migration within the home changed the use and meanings of spaces such as bedrooms, just as media migration to the house changed the use and meaning of the house as a home. The connected home is a node in a global network, as is the networked bedroom. Neither the home nor the bedroom is a sanctuary from online work, consumption, education, or entertainment. Efforts to create seclusion zones within the home were increasingly made—sometimes applying to certain spaces ("no screens at the dining room table"), sometimes to certain times ("no media before breakfast"), sometimes to certain people ("the children"), and often to combinations of spaces, times, and people, in often idiosyncratic entanglements of householders and media practices ("no media in the children's bedrooms after 10 p.m.").

Nevertheless, the steady accumulation of devices in the home through this time began to push against the social values, norms, and modes of dwelling associated with technology-free zones. For example, as households upgraded their televisions to digital flat screens, obsolete models migrated to other places in the house—often to children's bedrooms. While in some instances this initiated processes of distributed viewing and critiques of "living together separately," we also observed unexpected contradictions, with the newer and larger higher-definition flat screen drawing family members back to shared viewing. Sharing the same space with multiple screens used for multiple purposes was, however, not without its difficulties. The fact that wireless services enabled more flexible use of domestic spaces was not simply a smooth transition from bounded to dispersed home media use. As we encountered at this time, and discussed further in Chapter 4, families were negotiating these emerging dynamics of media and family life, having to navigate the ways media were and should be used by different family members in the home.

The kitchen, too, emerged as a contested space in which the appropriation of digital media had to find a place alongside other appliances. Cooking technologies such as a stove, sink, knives, and so on are of course essential to the configuration of a kitchen as a kitchen, but for a long time kitchens also included media in the form of cookbooks and magazine recipes, calendars and bulletin boards.

As one might expect, tablet screens and phone screens now compete with print media in this space. Recipes in text are now appropriated as YouTube video demonstrations, recipes are accompanied by selected music tracks, and digital media colonize the kitchen:

> The laptop mostly sits in the kitchen, the only time it moves is when the cleaner arrives to do the kitchen I pick it up and walk out with it . . . I often use it for recipes. I have two Google files with recipes. Like the other day I wanted to make a goat tagine, so I looked for a recipe on the internet, I use it for that sort of thing. I think it's quite handy. And it's out of the way. I've got a breakfast table that looks out on the garden and the breakfast table is big enough to have the laptop at one end while I have breakfast at the other. It suits me fine. (Odette, High-Speed Broadband 2011–2017)

Just as kitchens became mediated spaces of cooking, dining rooms, too, while used for eating were increasingly accompanied by a variety of screens and their streamed media. Bathrooms were used in parallel with speakers. Bedrooms became especially amenable to the use of mobile devices, yet against the affordances for using mobile devices in bed and the pull of checking social media feeds just before going to sleep or first thing on waking, old sentiments of privileging bedrooms for rest lingered:

> It [notebook] goes from the bedroom to the couch. [laughs] Well, it depends on the time of the year. In wintertime, it gets cold and I'd prefer to not heat the whole house, so I'll heat the bedroom, 'cause I have my big comfy chair and TV set and stereo and my bed and everything, so it's not as if I'm going off to a little dark corner. In summertime, it's really hot upstairs, and

downstairs is much cooler, so I'm not gonna watch TV or work on my computer upstairs. I don't usually take the computer to bed though, because it's sleep. (Dawn, High-Speed Broadband 2011–2017)

In addition to media migrations and accompanying changes to the use and meanings of those spaces, the historical designation of some rooms as "no-go" zones became either redundant or difficult to maintain. We nevertheless still saw forms of technology exclusion. Such efforts were, however, more often idiosyncratic or particular to situated entanglements of householders and media practices, and less applied to space type per se. Moreover, exclusion zones were as likely to be applied selectively to particular technologies rather than to all technologies in a specific space (Figure 3.3).

THE INFRASTRUCTURE OF SMART HOMES

The processes of diffusion, accommodation, and appropriation reached a peak with a vision of dwelling in a home filled with smart interconnected networks of devices to seamlessly control the environment and life within it. These ambitions emerged far earlier than today's Google, Amazon, and Microsoft home assistants, with some households in the Connected Homes project (2004–2010) attempting to configure their own version of a smart home long before the availability of these off-the-shelf, plug-and-play devices.

One family we met in the Connected Homes project (2004–2010) in particular, offered a glimpse into early smart home technology appropriation to control the security, lighting, heating, curtains, stereos, and televisions in their house in 2004–2005. There were many televisions in their home and in the outdoor living

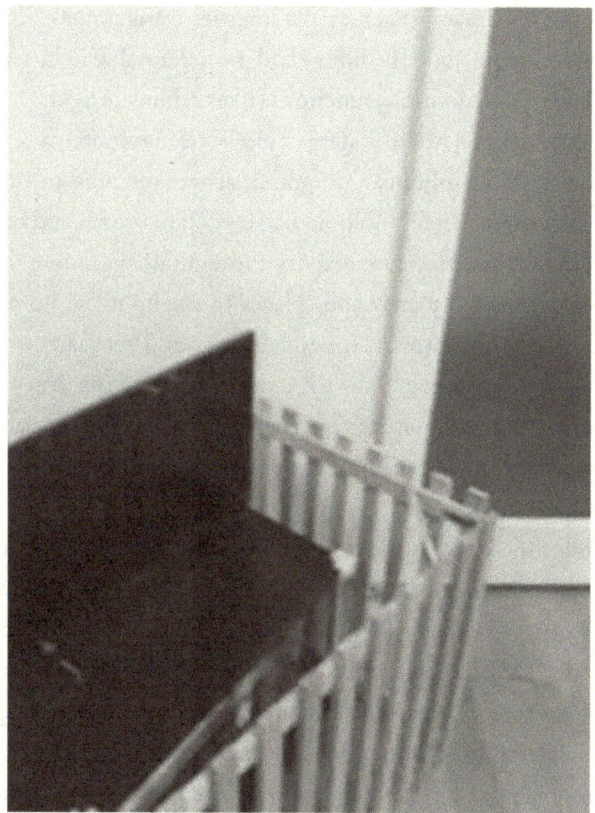

Figure 3.3. A toddler-proof fence excludes the TV from the living area (High-Speed Broadband 2011–2017)

areas, each of which could play cable content, DVDs, videos and free-to-air television, as well as images from the CCTV security cameras dotted around the house and gardens. The stereo music system was also centrally networked so it could play in various parts of the house, and a separate cinema room was equipped with a large-format high-definition projector. Curtains opened and closed via a remote-control handset. The kitchen had cable and ports for "network appliances" (such as the internet refrigerator), although none were connected. The house had programmable light switches that could perform various functions (like turning on lights!). While the basic functions of the system could be used remotely via the internet and mobile phone, a programmer was required to make all major alterations. The household had a satellite broadband connection, wireless internet connectivity throughout the home, as well as cable connectivity at various places throughout the home. The house was littered with screens and devices, and yet many were not in use or not connected to the network. Unusual at the time though this household was, perhaps even eccentric, the appropriation of contemporary media technologies for these and other purposes is now commonplace.

While the concept of the connected or smart digital home had been around in different guises for a number of decades (see, for example, Spigel 2001; Venkatesh 2008; Harper 2011), broadband infrastructure anticipated a much more productive digital home. Bandwidth became part of a wider infrastructural understanding and imaginary of the digital home that enabled the multiplication of devices, especially mobile and touchscreen devices. Nevertheless, we found that such ecologies and imaginaries also pushed against the architecture of domestic space, with the population and possibilities of mobile media living not necessarily mapping onto the design and layout of existing home spaces, leading to questions about not just how technologies were accommodated in domestic spaces, but also how the materiality of domestic architectures was able to (or not) accommodate new technologies, infrastructures, and their associated practices.

The move toward smarter devices, then, was bound up with perceived logics or accepted imperatives of upgrading and accumulating technology. This often combined with perceptions of the home internet infrastructure slowing and becoming insufficient for the evolving household media practices:

> He notices that they start to slow down when they become full of software, data, etc., it is worth it in terms of productivity. (Bob, Connected Homes 2004–2010)

> It [ADSL connection] was a pain in the neck, it would drop out and that sort of stuff. (Sharon, High-Speed Broadband 2011–2017)

The move toward smart devices was also part of wider visions and imaginaries of mediated living, including the myth of seamless connectivity, or the home as a node in a larger network of productivity:

> I get to have the NBN [fiber internet connection], which is awesomely fast. And that's kind of it. Because of my work I spend most days and a fair bit of the evenings on the internet stuffing about [not acting efficiently or purposefully], and I just want to be able to have access. I do a lot of remote access to my computer at work and I just want it to work. I just want to go, I need that file, grab, put, shift this, upload that, bang, I don't want to have to think about it. I always think about what Arthur C. Clarke said, "in the future when technology works well it will seem like magic." When I first started working with Macs, which I love, you'd always tweak them, and get your settings right and download a new version of something, and investigate all the different things it would do. And now I can't be bothered, I just want to plonk something on a machine and push a button, and I want it to go and I don't want to think about it. (Pat, High-Speed Broadband 2011–2017)

THE APPROPRIATION OF NATIONAL INFRASTRUCTURE BY THE HOUSEHOLD

The appropriation of the kinds of technologies we observed in the house of the early adopters from the Connected Homes project (2004–2010) was in turn dependent on the appropriation of precursor technologies, the most significant being the provision of high-speed broadband to the house. Many studies of media use in Australia echo a recurring theme in reports on media consumption: As the speed and availability of the internet increased, the appropriation of digital media services grew.

Following the global financial crisis of 2008, many countries around the world began investing in broadband infrastructure "to jump-start economies weakened by the recent financial collapse" (West 2010, 2). Like other "nation building" infrastructure projects such as road, rail, and utilities like electricity and water, broadband was seen as an essential investment in the country's future—able to act as a short-term stimulus to boost economic growth by providing jobs through its design and implementation. In Australia, this project was captured under the heading of a national broadband network (NBN).

The NBN was originally conceived as one of a number of stimulus measures to prevent Australia from going into recession as a result of the financial crisis and was consistently promoted as a measure to provide jobs for Australian workers (Conroy 2009). This dual emphasis on the capacity of broadband investment to "stimulate jobs in the short-term and pay a dividend ... through enhanced productivity and innovation in the long-term" (Conroy 2009, n.p.) similarly underpinned policies adopted by governments and policymakers around the world investing in the digital economy. In this vision, the digital home was imagined as an integral part of a national and global network that would not only change the relationships of homes to workplaces, but also provide better opportunities for flexible management of work/life balance, better health care and education, and better access to government services; overcome geographic barriers and the tyranny of distance for economic participation and productivity in regional areas; and reduce commuting and the associated impacts on congestion, fuel consumption, and carbon emissions on the environment (Department of Broadband, Communications and the Digital Economy [DBCDE] 2010). Broadband was thus appropriated at a national level to mediate new systems that were evident at the macroeconomic level.

We found that this national logic of appropriation was consistent in some but not all respects of broadband appropriation logic at the household level. It was anticipated, for example, that the national appropriation of broadband would deliver growth in economic productivity (Conroy/DBCDE 2009), which would involve a fundamental change in home-based remote work practices (DBCDE 2010). Before the NBN, householders had not appropriated broadband for this purpose and telework was relatively uncommon in Australia (Australian Bureau of Statistics 2009; DBCDE 2010). In its 2010 report on teleworking, Access Economics reported that only 6 percent of Australian workers had a teleworking arrangement with their employer, compared with 10 to 11 percent internationally (Access Economics 2010, i). By 2016, however, this figure had risen to 30 percent, or 3.5 million people from a total workforce of 11.6 million (MyNetFone 2016).

> My husband works from home. Not always, but he has a home office. He is a contractor and does a lot of work from home. He connects remotely to an off-site computer. So, actually, that is another reason [for getting the NBN]. It's faster and less frustrating

for him, when you are connecting remotely there was often a bit of a lag and there is less so now, which is more convenient for him. He does a lot of reports, which will be on the remote computer, and it used to be a bit slow, like to edit something would be a bit frustrating, the lag, but he doesn't get that anymore. (Sada, High-Speed Broadband 2011–2017)

Critics of the NBN in its fiber-to-the-premises form suggested that it was a waste of money to spend billions to enable householders to download "cute-cat videos" faster, and, indeed, video streaming is becoming a significantly more vital part of internet use, with the DBCDE (the Australian government department that managed communications infrastructure at this time, now called the Department of Communications and the Arts) underscoring that "video communications will be an important driver of bandwidth usage in the years ahead, particularly high-definition video" (DBCDE 2011b, 93). As Phil Ruthven (2012, 32) notes, internet video data were, at that time, estimated to account for more than 50 percent of consumer internet traffic. Looking ahead, Ruthven predicted that "if we assume a 40 percent growth in data annually over the next decade, this would see Australian consumers requiring a monthly data allowance approaching 200GB by 2020." In addition, if these figures were extrapolated over the next forty years,

> it is not hard to imagine broadband networks needing to support individual consumer monthly download limits in excess of 5TB by 2030. Faster broadband networks are required to support these downloads, and by 2050 we may require broadband speeds in excess of 10 Gbps. (Ruthven 2012, 32)

In terms of the day-to-day use of the internet, high-speed broadband was expected to significantly boost both the amount of content being consumed at home and the speed and efficiency with which it was accessed through various platforms.

As one might expect, we found that different households appropriated broadband in different ways, with some simply maintaining limited and quite prosaic uses, such as paying bills and checking emails. The household's expectation was that the NBN would do what the internet had always done, but faster and more reliably, and it was not imagined as a technology that would fundamentally reconfigure household media practices.

> I probably have enough computing power on my computer to design the space shuttle and I use it 90% of the time as a word processing machine. And the speed of the internet I could probably do something extraordinary but I won't use it . . . (Grant, High-Speed Broadband 2011–2017)

Others, though, became voracious media technology consumers and users, often bound up with more consumer-oriented and aesthetic reasons associated with being early adopters of new technology:

> Bob likes to keep up to date with the best technology, he updates every 18 months. (Katie, Connected Homes 2004–2010)

> If there's a whole new jump in technology I'll want it . . . I like the extra functions that you can get. I am quite excited at the moment by the fact that I can remotely log into a server that is sitting at my desk at my office. Like I can actually get my desktop up from my computer at the office, the whole desktop, even the screen saver . . . and look at it here at home, and that's just

crazy. Those sort of things I think are really kind of groovy, and I reckon that next step up with TV (IPTV) I would be interested in ... the library would just be sitting there ... and you'd just go click and watch it. (Tom, Connected Homes 2004–2010)

I love gadgets ... my laptop is quite heavy, what I want is something light. I don't like carrying heavy stuff. That's part of the reason why mobile phones and iPods and those things annoy me. I'd love to have everything all at once. I like the look of the iPhone, but I won't go down that track, it's just silly, 'cause I've got a phone and I don't use it very much. But the iPad, I like the idea that it's little and light, and it will do the sorts of things that I want to do. (Georgie, Connected Homes 2004–2010)

The appropriation of multiple devices and the ambient presence and circulation of mobile media were discussed by families in terms of the rationale for and patterns of household mobile media adoption, and these forms of appropriation often revolved around the perceived needs or benefits for children, articulated in terms of their suitability for use by children. Tablet devices, in particular, were perceived to offer affordances for children associated with usability, mobility, and multifunctionality. In such contexts, of mobile devices and screens, along with wireless connectivity throughout the home, we saw new and novel practices emerges, such as Skype babysitting:

We have weekly times when we call on Skype. We're starting to use with the kids. Now with Sophia, we can stick her on and they [grandparents] can see her ... when she was a bit younger she was in the Bumbo and I got mum to watch her for a minute while I walked away to light the fire. (Natasha, High-Speed Broadband 2011–2017)

We also found a relationship between household composition and appropriation of increased bandwidth. Early adopters of the NBN were more likely to be households with children than those without children, more likely to have higher incomes, and much more likely to be households who own their home rather than rent (Nansen et al. 2013a). The households that had taken up an NBN plan were also more knowledgeable consumers of broadband technologies, suggesting that participants tended to reflect the characteristics attributed to early adopters by the diffusion model. We found that households that had appropriated the NBN were more likely than other fixed-line broadband households to have a convergent communications ecology, to use their home for business or telework, and to engage in a greater variety of online activities. For these homes, increased bandwidth was accompanied by increased use of digital media in the home for entertainment and communication purposes. These homes were already media-saturated environments (e.g., Australian Communications and Media Authority 2007), though clearly, by today's standards, much less so.

A significant factor compounding the logistics of appropriation by households was the integration of the NBN internet connection with existing digital infrastructure and with each household's domestic ecology of hardware devices, internal routers, software, and of course the household's skill and interest, which was not always a seamless arrangement (Figure 3.4). Thus, while vastly increased bandwidth was the key technical affordance to be appropriated, connecting fiber to the household did not automatically raise connection speeds to advertised rates. Instead, download and upload speeds were influenced by a whole raft of household factors, including Wi-Fi as compared to cable; the age and capacity of modems; the number and position of desktops, laptops, and so on;

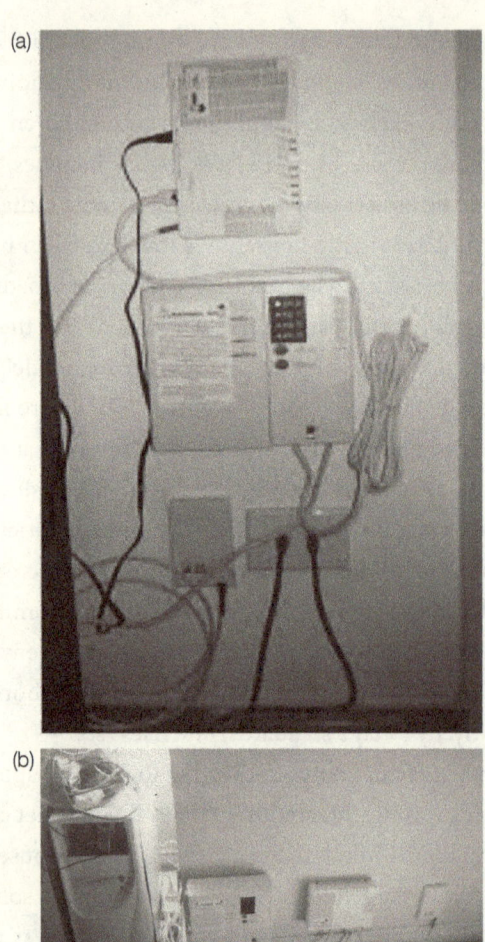
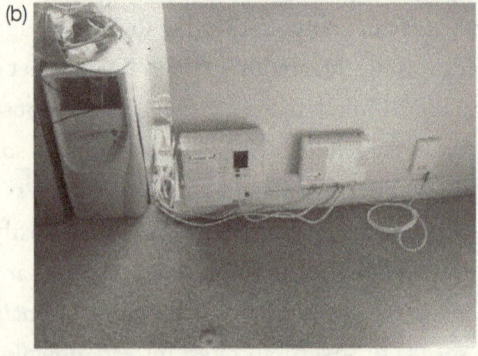

Figure 3.4. Photograph of installed NBN household connection hardware (High-Speed Broadband 2011–2017)

the number of simultaneous users; the layout and material composition of the house; and the competency of householders to set up or manage a local area network or the interoperability of devices (as we discuss in the next chapter):

> The wireless router is down there [downstairs office], and if I sit on this end of the lounge it's okay, but if I sit on that end it drops out . . . it's really frustrating . . . it drops out and you lose everything. (Michael, High-Speed Broadband 2011–2017)

The national ambition that households would accommodate the NBN's capacity to provide fast, universal, and equitable digital access was thus challenged by internal household media ecologies, even within a single geographic location. Important in these media ecologies were the demands the technologies made, which could often only be accommodated by the presence of still more and still newer technologies:

> We had to buy a new router. Because we were on a 100 megabit plan, the WAN to LAN throughput speed expected by most ADSL routers is 24 megabit down speed, so we needed to upgrade that. So, I did a bit of research and ended up with an Apple Airport Extreme—it's the only Apple product we use—specifically because it has a really fast processor to handle this. It was the cheapest way to get a router that supported these speeds, and we got rid of our ADSL modem that we were using. We purchased that separately. iiNet were gonna offer us some routers and they had a bunch of different ones but we didn't really want any of their stuff. (Andy, High-Speed Broadband 2011–2017)

As part of the entangled ecologies of household media are the visible devices in the home, but also their relation to the often invisible or hidden infrastructure of household internet, which combine to help construct these environments (Mackenzie 2010). So, for example, a wireless router connects to the power supply, the aerial, the circuit board, and so on. The wireless router is itself a component of the home network, which might itself be a component of a smart home communication and control system, which is in turn a component of the broadband network and then the internet. Sometimes the materiality of such interrelationships did not neatly align, and we found instances in which the furniture or fittings of homes encumbered the connection to the internet:

> When we got it [the NBN box] put in here, they put the cable over and it got caught up in our blinds and the blinds made it stop working. We've got external security shutters. We don't know where or how they put the cable, but they put the cabling up through the mechanism. There must have been a gap where they just pushed it through. The cable they put in was hanging down. We contacted them to fix it and now the blinds don't work. (Peter, High-Speed Broadband 2011–2017)

What we saw here, then, were considerations about infrastructural placement that were driven by the materiality of technology, such as connection to the router and then to other broadband-connected devices. Without these fundamental relations, the internet disengaged, and the aggregate object we call the home network fell apart.

This chapter has shown how over the course of our studies, homes were incrementally populated with new technologies, each with needs to be accommodated, and with capacities to be appropriated in a variety of ways. The steady colonization by new technologies of the household media ecology, from their beachhead in the home office, to the living room, to the kitchen, and then to the bedroom, was associated with changes best described as ecological, to capture the interrelatedness of the presence and performance of the home's various human and technical inhabitants.

The significant changes in the ecology included the replacement of desktop computers with laptop computers then augmented with still more mobile pads, the replacement of analog televisions with digital flat-screen television, the replacement of wired devices with wireless devices, the smart home as an imaginary, the appropriation of high-speed broadband through the NBN, and the proliferation of increasingly sophisticated mobile phones. As one would expect, these changes to media and infrastructure—and their affordances and appropriations—had important implications for dwelling in the home. In an exercise of rational, planned appropriation, some of these implications were anticipated and actively assessed (such as the positioning of the television set) but less so in other cases (such as the move of mobile devices into bedrooms and other hitherto "no-go zones"). Where and when communication, entertainment, or work was undertaken within the home, and the distribution of functions across various devices, was unsettled, with implications for dwelling. Further, previously differentiated activities and resources—for work, for play; for adults, for children; for the living room, for the study; for the office, for the home—became increasingly mobilized and distributed, dedifferentiated and mingled, and challenged previously idealized notions and aspirations couched in terms of binary socio-technological and spatiotemporal arrangements of dwelling.

Changes at the level of media materiality are complicated, and imagined synchronization of occupants and infrastructures remains elusive. Dwelling with broadband-fueled wireless, distributed, and mobile media practices made dedicated domestic space allocation and time allocation redundant and participated in the constitution of new modes of dwelling.

Our participants have, over the last seventeen years, helped us to reflect on the difficulty in maintaining the legibility and coherence of digital technologies through the accumulation, accommodation, and appropriation of devices in their homes. Unsurprisingly, families reported on the fact that the volume of their media use increased, both in terms of the range of applications accessed and in terms of the amount of time, and how connected infrastructures and mobile devices meant denser relations of mediated communication and content consumption. We see domestic media ecologies that are, if anything, overpopulated with screens of different ages, devices with overlapping functionality, technologies in different states of working order, and applications and operating platforms jostling for position on devices and fiercely competing for the attention of the householders and other devices—the modem, the large flat-screen TV. Often these new technologies appeared to be accumulated by default, and the accumulation of new devices and services was accepted as part-and-parcel of contemporary living. Short life cycles, quick turnover, and high redundancy rates would appear to be a digital industry standard, and life for a device in the household ecology is often short and brutal as householders struggle to accommodate the demands of technologies and appropriate their functions in ways that smoothly integrate with the ecology as a whole. In the following chapters we turn to these messy and entangled dynamics of managing and negotiating household media ecologies and, in turn, to the ecological issues of consumption and obsolescence in discontinuing and discarding technologies.

[4]

HOUSEKEEPINGS

INTRODUCTION

This chapter moves from the focus of previous chapters on media adoption, appropriation, and accumulation to focus on the significant and often invisible forms of maintenance work needed to keep things functioning and fit-for-purpose in the domestic media ecology. It shifts from an historical register to a more contemporaneous consideration of the household labor involved in investigating options for, making decisions about, and purchasing and setting up new technologies as well as their ongoing maintenance. The chapter examines the work, and who does the work, of maintaining and managing digital media. It also explores the relations of power, authority, gender, labor, and expertise that go into making decisions about and maintaining and using household digital technologies, and, in doing so, furthers empirical developments concerning the notion of domestic media ecologies. Consistent with our theoretical approach, we treat the subjects and the objects of these labors somewhat symmetrically, looking not only to the work done by the householders to build and maintain a functioning media ecology but also to the way this work folds back from the technology to build and maintain the identity of householders.

Digital Domesticity. Jenny Kennedy, Michael Arnold, Martin Gibbs, Bjorn Nansen, and Rowan Wilken, Oxford University Press (2020). © Oxford University Press.
DOI: 10.1093/oso/9780190905781.001.0001

At the turn of the century it was uncontroversial to assert that domestic technologies were gendered in their design, construction, patterns of use, and ideology, and that technologies were similarly discriminating in terms of age or generation. Today Cynthia Cockburn and Susan Ormrod's (1993) "white-goods"/"brown-goods" gender binary and Prensky's (2001) "digital natives"/"digital immigrants" generational binary are not nearly as crudely drawn, but it remains the case that in any particular context, technical work done in the household and affective responses to that work are not evenly distributed; consequently, relations to technologies constitute a resource for identity construction. While differences within and between households often arise from personal preference, these preferences are often stabilized over time to help define the household's settled division of labor, and one's position in that labor market folds back to articulate a personal identity. Interestingly, identity markers that might apply to both the person and the technology, such as "useful," "contemporary," "reliable," and "smart" (or their antonyms), are frequently employed. In this chapter, therefore, we argue that relations between individual householders and the labor associated with the maintenance and use of media technologies constitute important fields for the achievement and designation of identity (some would say ontology)—for both the ecology and the householder.

While sometimes imagined as reducing labor in the home (see: Dourish and Bell 2011; Gregg 2018), new technologies often require significant work in this ongoing process, and this includes digital media technologies. This finding is supported by research, including Tolmie et al. (2007, 332), who use the term "digital housekeeping" to refer to "the work involved in setting up and maintaining home networks." Based on these arguments, in this chapter we set out the kinds of work that constitute digital housekeeping; we examine the forms of digital expertise that are available and valued within the home; we look at how expertise is performed and distributed among household members; and we consider the conditions of labor under which such expertise is accessed or required, paying particular attention to the gendering of this labor. These aspects of work, expertise, and gender are situated in relation to the demands made by the evolving dynamics of household media ecologies, newly impacted by the appropriation of broadband infrastructure.

As part of this discussion, we consider how digital housekeeping also implicitly situates technology work within the home in the role of the "housekeeper," a term that is complicated by gendered sensitivities to the distribution of domestic labor (Blythe and Monk 2002; Bell et al. 2005). Already established in feminist-oriented research is a consideration of how the identities that are negotiated in relation to technology are gendered, and how particular technologies are themselves gendered in their uses and perceived meanings (Berg 1997; Bergman and van Zoonen 1999). A feminist approach draws attention to such labors and shows how expertise is unevenly acquired and the ways that values are inscribed in constructions of expertise. Furthermore, such an approach draws attention to the ways that expertise is motivated by particular desires and interests.

The findings presented in this chapter indicate how transformations within household media configurations give rise to new opportunities for "digital expertise" under conditions of persistent gender inequalities. The literature on expertise identifies a range of strategies, activities, and characteristics. These can be loosely summarized as agency in decision making; the ability to see meaning and implications of actions; the interiorization of cognitive capacity; and a sense of responsibility (e.g., Glaser and Chi 1988; Shanteau 1992; Cellier 1997; Farrington-Darby and Wilson 2006). Specifically, new media debates locate expertise in discussion of "literacy" and "digital natives," indicating it as an inherent characteristic that omits the labor of becoming expert. We identify how the expertise that is developed in the home is applied to other social fields and consider the cost of maintaining a networked home in terms of the labor required and its tradeoffs against other household activities. The work of situating and maintaining technology in the home emphasizes the social and material construction of technology in everyday life.

In the case study that follows, we see that many of the issues identified above were rehearsed as a family from the Connected Homes project (2004–2010) managed digital content and their domestic digital infrastructure, as they distributed the digital housekeeping tasks among themselves, as they exercised varying levels of expertise and interest, as all of this is shaped by gender, and as gender is shaped in its interactions with digital housekeeping.

A CASE STUDY OF GENDER AT HOME

John and Mary, both in their mid-forties when we met in the mid-2000s, lived in the inner city with their two children (Connected Homes 2004–2010). Peter, their firstborn, had just turned twenty, and eleven years his junior is his sister Angela, who was about to have her ninth birthday. Their house was a semidetached single-fronted weatherboard that began its life as a worker's cottage a hundred years ago. Like most houses throughout this neighborhood, it had been extensively renovated to accommodate new generations of more affluent owners. Also like most houses in the area, this one was well equipped with the most current devices at the time, including several networked computers, broadband internet connection, television sets, a DVD player, stereo equipment, fixed-line telephones, and so on. Historically, this had been a working-class area but, reflecting the increasing overlay of a middle-class demographic in this area, John and Mary were tertiary-educated information workers. John was employed in an IT role for a large bank, and Mary, having left teaching, was now a publishing editor for an education firm. Peter was a tertiary student with part-time work at a music shop. Angela attended the same local primary school that Peter had attended a decade ago.

After repeated visits over a five-year period, it appeared that all members of the family felt comfortable with digital devices. The family seemed adept at using digital technologies to meet their various needs for communication, entertainment, working from home, and so on. Making and receiving telephone calls, searching the internet, operating the stereo or the DVD player, tuning in to radio programs, and watching TV were all part of the routine of home life and family life. This combination of comfort, familiarity, and skill in using digital technologies, however, was underpinned by subtler dimensions of relationships and identities that inhered in both perceptions of self and others and attitudes toward digital technologies. In John and Mary's family, these impinged on understandings of what communication with others was about, what digital technologies were for, and what particular forms of social life were desirable—within the family as well as between the family and the "outside world."

Yet we couldn't adequately appreciate the complexity of the dynamic between householders and digital technologies by treating the household as a single homogenous unit. Family members had markedly differentiated perceptions of relationships and digital technologies. While difference might at times be a consequence of personal preference, these preferences were stable for long enough to settle into the various roles that family members assumed. We thus begin with individual portraits of John and Mary's technology use based on their own understandings and reasons pertaining to their technology use and their modes of sociality and conviviality (see Chapter 2).

Like most people in his social position, John routinely used the telephone and to a lesser extent email to maintain personal relationships with significant others. John telephoned to invite friends to a BBQ or to confirm attendance at his now notorious back-garden working bees. Within this context, what, for John, was the significance of communication? As he put it:

> There always has to be another reason to initiate the call. Like [the reason for] the call is to catch up, but it's more than just to have a chat, "let's organise something!" I really like my interaction ... to be structured around something ... something face-to-face and something more than sitting around talking. You do something ... share activity and take whatever conversation comes ... (Connected Homes 2004–2010)

If John saw his use of domestic digital technologies as facilitating communication for a particular purpose, he was nevertheless happy to engage in a telephone conversation, providing, it seems, that the ulterior pragmatics had been identified. John felt disconcerted by communication as "an end in itself," and by using digital technologies to structure activity, he reflexively stamped his identity as a man of action.

Mary was somewhat less comfortable with the telephone than John and in her research diary wrote she found it "quite intrusive." "You don't plan for it to ring," she said, resentful of others' ability to use the phone to "intrude on my peace of mind" (Research diary, Connected Homes 2004–2010). Consistent with this, and like

John, Mary rarely phoned just for a chat. More often, she used the phone in a task-oriented way when, for example, "there is content to communicate quickly" such as "who is driving who to tennis" or "to organise a table of friends for a school trivia night." Both Mary and John constructed digital technologies as functional devices oriented to the management of activity, and their use of technologies constructed and announced John and Mary as well-organized and active participants in social life. Mary's pragmatism extended to email, where she would use it "to arrange for a face-to-face social activity." But, again, it was this pragmatic quality that stood out. Where John took considerable care to craft his emails, no such aesthetic or literary concern was discernable for Mary. Indeed, in many respects, Mary's aesthetic values and her approach to digital technologies had few points of intersection.

Curiously, John was less inclined to capture significant people in his personal history than he was to capture the technologies in his personal history. In the parents' bedroom, where there were no digital technologies, John kept his much-loved childhood electronics kit on top of the wardrobe. Of greater significance was an array of devices that lined a shelf of the study, including an antique radio, a shortwave radio, an old oscilloscope, and a Morse code transmitter and receiver. As a former member of Amateur Electronics, John fondly recounted the history of these items. He also looked forward to repairing these items, particularly the Morse code transmitter and the shortwave radio. Just as a row of stuffed moose heads in a living room might attest to the victories of the hunter, John's displayed "museum pieces" identified him as a man who harbored a fascination for the objects themselves, and a man competent to make his way through complex technological systems.

Competence, however, has its limits. On several occasions, John expressed his anger for having overlooked the regional specificities of a DVD player when purchasing the item. As Mary reported in the scrapbook:

> John is . . . annoyed with himself, and feeling a little despondent . . . He brought the wrong thing-a-ma-jiggy DVD speaker thing. Hmm. He thought he did the research but forgot to ask one essential question about the product. The question was . . . can this thing play DVDs from all regions? Answer . . . alas . . . NO! The kid (the big one) has an extensive DVD collection from across the planet . . . Oops! (Scrapbook, Connected Homes 2004–2010)

Notwithstanding Mary's playful but perhaps ironic relegation of John's manhood to the category of "big boy-hood," John succeeded in resurrecting his identity as a figure of technical expertise. When the micro hi-fi DVD home theater system was all connected, John discovered that the DVD player was in fact region free, despite advice to the contrary in the instruction manual. John was finally very pleased with the system and delighted in an exhaustive visual rendering of "Instructions" in the scrapbook in an attempt to make it clear how to work the complicated connections between TV, DVD, VCR, CD, radio, amplifier, and speakers (Figure 4.1). Indeed, this was a source of real pleasure for him. "I'm rapt," he declared, while worried about "[Mary] having to learn a new set of technology. She knows she can do it, but that it will take time to learn" (John, Connected Homes 2004–2010).

In a relative sense, John's technological competence was co-constituted by Mary's lack of affinity with technology. At the same time, John's rapture seemed to correspond to Mary's indifference. Mary did not feel excited about new technologies, instead feeling somewhat removed from the stimulation

Figure 4.1. John showing his home entertainment system configuration (High-Speed Broadband 2011–2017)

that others in the family experienced in acquiring new technology (Research diary, Connected Homes 2004–2010). John tried to engage her in making decisions regarding the technologies, but it just wasn't her thing. Completing a sticky label from the "domestic probe pack" (see Chapter 2) and placing it in her probe pack diary, Mary admitted, "I feel disconnected to others when . . . we buy new stuff and everyone's excited and I just think, "oh yeah, that's good!" (Research diary, Connected Homes 2004–2010). While possibly compensating for a sense of distance at her lack of interest, Mary saw fit to point out that "disconnected" does not mean "discontented." But, on the purchase of the TV/DVD/hi-fi assemblage, Mary acknowledged that although John was "very consultable and democratic and all that . . . , sometimes I just have to say, 'I don't think I have much input here . . .' and that's perfectly fine. Really!" (Research diary, Connected Homes 2004–2010).

It is possible that Mary may even have felt a subtle call to technical disconnection. It is a realm that can be given over to John, and Mary was thus disburdened, whilst affirming her husband's identity as a technically competent man. When she tried to work out how to load and operate their camera, she wondered, "How long should it take to find the shutter button?" (Scrapbook, Connected Homes 2004–2010).

TECHNOLOGY COMPETENCE AND IDENTITY

As a form of social being, John's gender was a product of social structures and personal enculturation, and a wide variety of sociocultural and material resources were pressed into the service of constructing and maintaining his gender identity (Connell 1987; Rakow 1992; Wajcman 1994; Lerman et al. 2003; Wajcman 2004; Connell 2005). Among these resources were technological artifacts, affective responses to technologies of various kinds, and competencies in the use of technologies of various kinds. In any particular sociocultural context, technical competence and norms of affective response in relation to a particular technology might unify a gender identity (John was drawn to electronics and understood digital technologies at a fundamental level) and marked a point of difference (Mary was not, and did not). Elements of John's (and Mary's) identity became visible through and within John's interactions with people and with digital technologies.

Identity, however, is more than the role one plays in social life. Identity impinges on the type of person one wants to be, the kinds of projections that one makes in social interaction with others, and the kind of person one becomes or is. Moreover, identity is affective. John's identity was immersed in an emotionally dynamic vision of self as a technically competent man, a technical provider. As a technical provider, moreover, John was able to affirm his masculine identity around the counterpoint of Mary's feminine identity through their respective relationships with technologies. John compiled instructions to facilitate Mary's use of the home theater system, and he reassured her that it would just be a matter of time before she too learned to use it. Murray (1993, 64) maintains that "wherever [masculinity] struts around, it does so in a different relational dance with femininity," and technology is both a point of reference and an agent in this relational dance.

It is now uncontroversial to assert that domestic technologies are gendered in design, construction, patterns of use, and ideology (Wajcman 1994). While some domestic technologies are shaped within a dominant construct of female domesticity and labor, other domestic technologies are masculinized and located squarely in the male domain of socially defined interest and expertise (Rakow 1992). This division is commonly characterized as that which exists between white goods and brown goods (Cockburn and Ormrod 1993, 98). It would appear that in Mary and John's household, at least, electronic technologies were very much "brown goods," and in John's domain of interest and expertise. The household, in this respect, confirmed a frequently rehearsed model of gendered relations to digital technologies.

This apparently gendered enthusiasm on the one hand and indifference on the other may well have its origins in our experience of gendered socialization and enculturation—boys are brought up with toy machines and girls with toy people. In the context of this case study, an interesting argument is made by Kristen Haring (2003) in her work on early ham radio enthusiasts. Using the case of ham radio, Haring suggests that a factor in gendered responses to technologies is the effect that masculinized technology has in displacing women as the anchor point around which masculinity is formed. That is, in significant part, men are constructed as "men" through their relation to certain technologies, not only through their relation to women. Technology is the "other," not sex. Technologies, such as custom-made home theater systems, oscilloscopes, and Morse code transmitters, provide men, such as John, with points of self-reference that are ontological markers of their subject identity. In assembling the home theater or repairing the Morse code transmitter, John was constructing himself in a manner of self-reflexivity accessible to others. And if technical competence, in combination with the efficiency of the technologies as working artifacts, points toward a masculine identity (in John) that is robust and functional, the solidity of this identity becomes vulnerable when technology is breaking down or when it defies the efforts of the technician to work it. In this context, it is not surprising that John was "despondent" and "annoyed with himself" when he thought he had bought the wrong type of DVD player for the system and was subsequently very relieved to find that this was not the case after all (Connected Homes 2004–2010). Self-construction and identity were at stake and were occurring in association with a DVD player and international measures to protect copyright, not in association with Mary. Her polite indifference was neither here nor there.

John's technical-identity performance was also on display throughout the house. In the bedroom, it manifested through his electronics kit; in the study, through the networked computers and the "museum pieces" on the shelf; in the kitchen/dining room, through the home theater system; and through the system that enabled his son Peter to pipe his music through the house. Many people collect this sort of "precious junk" (Csikszentmihalyi and Rochberg-Halton 1981)—junk insomuch as it doesn't work and no longer has use value or exchange value, but precious as historically obdurate ontological markers. The antique radio, Morse code transmitter, and other artifacts marked where John had come from, what sort of man he is/was—they provided security and stability—but these subject–object relations also attested to change. John had moved beyond his amateur electronics, in terms of personal maturation and sophistication, skill and understanding, and professionally and socioeconomically. The artifacts attested to this, to who he was, and how far he had come.

And, of course, the artifacts that do this work were not any old artifacts. John might have had a collection of old wine bottles, or books, pottery, suits, or cars, but, importantly, he had not. Contemporary economics and culture provide people such as John with an overabundance of "given" objects that might be appropriated and used to populate a personal world—far more than can be consumed in a lifetime. From this vast extant world of potential cultural resources, we choose which to "domesticate" and which will remain estranged and wild. John has domesticated electronic technologies. Not only did he possess them, but his knowledge and experience tamed them. His ability to tinker with them, customize them, make them act in the world made them truly his. At the same time of course, he was theirs; the subject–object relationship had fashioned John since his boyhood experiences with Amateur Electronics, just as John had fashioned the gadgets.

Cockburn and Fürst-Dilić (1994, 16) note that, while men and masculinity are framed as technological, women and femininity are produced as domestic. They continue:

> Other qualities and values are mapped onto this: technology is exciting, progressive, of high value. Domesticity, by contrast, is humdrum, repetitive, and low value. This is what gender is: an *ordering* process. The gender–technology relation involves the

production and reproduction of a hierarchy, between women and men, the masculine and the feminine . . . the technical and the social. (Cockburn & Fürst-Dilić 1994, 16, emphasis in original)

While Mary's identity as a gendered subject was partially defined by an indifference to technology that complemented John's competence and enthusiasm, John's identity as a gendered subject was partially marked by an indifference to non-electronic forms of communications technology. Mary's collection of letters, for instance, did not evoke interest or enthusiasm in John, and when Mary produced her collection and spoke about it, John distanced himself and had nothing to contribute. The letters were clearly in Mary's realm of interest and expertise, and in distinguishing between the letters and digital technologies (and, thus, between herself and John), Mary was assertive in defending her position:

> I just can't think of anything in today's communication where their personalities would be captured. Looking a hundred years into the future it's hard to see how people would connect with our time. Writing is reflective—it offers depth. Blokes like that [soldiers] will write things that they would never verbalise—even though the letters weren't all that intimate. (Connected Homes 2004–2010)

The hierarchy that Cockburn and Fürst-Dilić and others point to is clearly contested. Mary did not accept that the social is subordinate to the technical, that domesticity is subordinate to technology, that the letters are subordinate to the oscilloscope. And there was a symmetry between Mary's relation to the letters and John's relation to his collection; neither had an interest in the other and each to their own. The objects thus provided these subjects with ontological "markers" that announced their differences while defining and shaping their subjectivity.

In her research diary and scrapbook as well as in conversation with us about letters and writing, Mary used the word "capture" repeatedly (Research diary, Connected Homes 2004–2010). For Mary the written word would seem to have an affordance of affective intransient intimacy not present in other media. While her own letter writing and receiving was infrequent, letters were important in capturing intimacies, capturing a young person's first experience of a foreign land, capturing other generations and war, and capturing a parallel life led on the other side of the world. These "capturings" had profoundly emotional implications for Mary.

Her emphasis on "capturing" could only be appreciated relationally as a counterpoint to the negative affordance of "loss" that modern communication technology offered. The letters were clearly objects that provided Mary with powerful points of self-reference. Like John's electronic gadgets, they marked who she is/was and where she had come from. The ability to capture experience and hold it fast provided the comfort and security of obduracy and stability. Whereas John chose to make electronic transmitters and receivers his own, Mary had chosen to lay claim to letters and populate her internal and external world with this collection.

Both letters and electronic gadgets act in the world, and this action is constructive of self and expressive of self. Mary was quite clear that letters acted in a way that electronics did not. Letters "captured" reflections. On the other hand, electronic technologies transmitted and received signals, and while his antiques no longer functioned, John had numerous contemporary devices that did, and these acted in the world.

John was more than just a talker: He was a doer, a social organizer, a role that was achieved through doing. Accordingly, John's style of relationship formation and maintenance might be seen as a chronology of "social events" in which interpersonal communication was a byproduct of the desire to "do." As we have seen in the preceding section, John understood this in terms of the difference between "just chatting" and "doing something." John acted in the world through the telephone and through email. His telephone use was purposeful, not phatic. His emails were carefully crafted and he was conscious of the semipermanent record of speech-acts that are implicit in each. To email was not to chat; it was to act.

John's approach was consistent with the view (given public prominence by Naomi Wolf) that boys (and then men) often bond around tasks—shared activities in which achieving, or even "winning," is valorized as the purpose of being together, whereas girls (and then women) bond around conversation and shared confidences, in which bonding itself is valorized, and affection is the prize (Wolf 1990). Should this be so, it might be noted that its implications for technologically mediated relationship formation are problematic, at least for Wolf's vision of men. Social interaction at a distance can certainly be read as more akin to a form of "talking" at a distance than a form of "doing." While John's employment was certainly an electronic form of "doing" at a distance, and while an electronically mediated form of recreational "doing"—an electronic equivalent of John's favored bush walk, or backyard working bee, might emerge—it was not in John's life yet. For John, digital technologies facilitated the doing and, where possible, John used domestic communications as a means to an end—as a method to organize, coordinate, and get things done. It was not an end in itself but acted materially in the world in a way that was constructive of self and expressive of being in the world, and of self.

But, in this respect, the situation in the household did not simply reinforce the claim that women like to "natter" whereas men like to "do" (Spender 1995). Mary reported that, while John was quite happy to talk on the phone, she found the telephone "quite intrusive" (Research diary, Connected Homes 2004–2010). As we have seen, Mary, like John, preferred to use the phone in a task-oriented way. Mary was also a "doer" who uses digital technologies strategically, in an ends-driven way. The examples she provided (i.e., to organize social events), however, recalled the results of a prominent North American study in which it was shown that when women made functional calls, they mainly served a collective interest. For men, in contrast, functional telephone traffic served mainly personal interests. These results showed that, in modern telephone culture, gender continued to mark significant differences (Frissen 1995, 87). The couple, though, did not reflect this difference. Both were pragmatic, task-oriented telephone users, and it was John who was comfortable to chat around a task, not Mary. John's use of digital technologies in a recreational context served social purposes—organizing groups for bushwalking trips, working bees, backyard BBQs—not simply private interests. Mary's use of email was minimalist-functionary, whereas John was more inclined to formal expression. In both cases, the telephone was used by both John and Mary to orient the family to the world and integrate internal affairs with external affairs.

PRACTICES OF DIGITAL HOUSEKEEPING

While early research on technology in the home often focused on moments of acquisition, including the preacquisition period where the technology's place in the existing ecology of the home was imagined, then the early processes of mutual accommodation, in which technologies and householders are physically and symbolically located in relation to one another in the home, it is important to note that this mutual accommodation remains an ongoing process well after acquisition (Haddon 2004; Miller 2012). As part of the ongoing process of mutual accommodation in household media ecologies, the formation and negotiation of social "rules" around how household members interact with particular technologies are important. Household members, for example, may exert control over others' access to or use of digital media in the home, based on issues such as location, ownership, or the division of labor. And efforts at accommodation may be resisted by the technologies themselves, by existing rhythms within the household, and by other household members.

Housekeeping activities serve to maintain the functioning and maintenance of the household media ecology and are associated with a wide range of tasks, including meal planning and preparation; shopping; cleaning; laundry; maintenance and repairs; care of adults, children, and pets; management of bills and expenses; and transportation (Arrighi and Maume 2000; Cunningham 2007; Lachance-Grzela and Bouchard 2010). These forms of work are usually conceptualized as unpaid and may be conducted routinely or intermittently (Lachance-Grzela and Bouchard 2010, 769). Australian census surveys documenting the distribution of these tasks within the home repeatedly identify gender division in the hours dedicated to performing these tasks (Australian Bureau of Statistics [ABS] 1997, 2006, 2009). Such surveys do not include tasks related to digital housekeeping, which this chapter positions as an additional form of (unpaid) labor in the networked home.

In what follows, we identify the types of labor digital housekeeping involves and discuss how they contribute to the organization and rhythms of media ecologies in the home. We begin with two of the more basic housekeeping tasks, the management of digital content and the management of digital networks, told through reference to several households and their varied practices. We follow this with a discussion of the expertise required to successfully complete this housekeeping. It will be seen here that variety rather than consistency is the theme.

MANAGING DIGITAL CONTENT

One of the most prevalent forms of digital housekeeping to emerge through our data was *access to and management of digital content*. Homes with high-speed broadband showed a strong preference for accessing digital content on demand compared to broader national patterns of internet use (Ewing and Thomas 2012). In this context, identifying, accessing, storing, and organizing digital content was a significant feature of digital housekeeping (Tolmie et al. 2007), performed in reaction to, or anticipation of, household needs. For instance, Doug lives in an area of Victoria, approximately four hours east of Melbourne, with his young family (High Speed Broadband 2004–2010). His family accessed high-speed broadband through a less-than-reliable satellite connection. At the time of our first interview, Doug was downloading a recently released animated movie for his family to enjoy later that evening. Their Saturday night movie routine was reliant on Doug identifying a suitable "family-friendly" movie, locating it online, and downloading it in time for them to watch. Negotiating with the family on the movie to download could be problematic, and downloading was also problematic with their intermittent satellite connection. If the service was interrupted it could take several hours, so Doug had to schedule this task ahead of time. Once the movie was located and downloaded, Doug stored the movie file on a two-terabyte hard drive, which was connected to a DVD player. The movie file remained on the hard drive after initial viewing in case a family member wanted to rewatch it at a later time. Through this exercise, technical skills were required, as were the skills required to negotiate family politics and interpersonal relations. In managing digital content, the father role and the IT role were entangled.

In Gloria's household, Emily, the middle daughter, was dubbed "the download queen" (High-Speed Broadband 2004–2010). Her siblings would come to her and ask what new movies she had downloaded or ask her to download a movie, TV series, or game for them, which they would then transfer onto their own device for individual consumption. For both Doug and Gloria's households, technical tasks in relation to content were chiefly concerned with content acquisition, amid problems associated with their limited broadband connection. For other participants, technical tasks were chiefly concerned with content storage and organization, because faster connections meant acquiring content was a less labored process. In each case, family members first needed to cooperate in the selection of content, and then the technology needed to cooperate to download that content. A visit to a home in the inner suburbs of Melbourne's north, shared by three young professionals, highlighted the variations in the degrees of labor required to access content. On a fiber-to-the-premises (FttP) connection, Christine, Alexis, and Shawn could download content very quickly: "It doesn't require much forward planning" (Christine, High-Speed Broadband 2004–2010). In a case of swings and roundabouts, though, what did require planning and attention in their house was the task of storing digital content. They had two hard drives, each with a capacity of one terabyte, connected directly to the television. One hard drive was for television series, the other for movies. When we first visited in 2013, Christine had recently reorganized the content on each of the hard drives so that it was possible for others to find it. The volume of content, together with the habit of dumping content in an operative folder (i.e., "downloads") in anticipation of immediate viewing, meant that the collection of files had become unwieldy. Prior to Christine's reorganization of the hard drives they found they were downloading duplicates of content because it wasn't immediately obvious that any given file had already been obtained.

Other participants described similar interpersonal and technical strategies for categorizing and organizing digital content. Riley and Ashley, an urban professional couple, had a home server for their digital content (High-Speed Broadband 2004–2010). Jeremy, living with his family in an outer northern suburb of Melbourne, also ordered his content on a central home server:

> It's very well organised. We don't delete things. I have a directory called movies. In there I have directory listings, there is a folder for any multiple movies like *Lord of the Rings* movies. If I need to shuffle things around to a different device like a laptop I can find it in the folders. (Jeremy, High-Speed Broadband 2004–2010)

Jeremy organized his media into "family-friendly" and "less family-friendly" categories. His control of content was also an exercise in technical skill and, perhaps more importantly for him, an exercise in fatherhood as he guided his younger daughter's growing independence in media consumption practices.

MANAGING DIGITAL NETWORKS

In addition to managing the acquisition and organization of media content, an important aspect of digital housekeeping is concerned with managing the materiality and the interoperability of the devices that store and present that content. The material form of the devices needs to be integrated into the domestic environment in aesthetically pleasing ways, as well as in a way that is functional. Aesthetic housekeeping tasks identified through the data include stacking hard drives neatly, hiding wires, making devices look orderly, and so on (Figure 4.2).

Furthermore, the work required to manage the aesthetics and the function of digital devices begins prior to the point of acquisition, when researching which particular technologies to purchase and investigating how they might be integrated into the current aesthetic and functional ecology of the home. For example, Carl, who ran an online marketing business from his inner urban Melbourne

Figure 4.2. Example of labor gone into the material management of household (High-Speed Broadband 2011–2017)

home, carefully researched interoperability between new and current devices prior to purchase—a nontrivial task that often occupied many hours (High-Speed Broadband 2004–2010). Carl's experience exemplified a number of themes that other participants in the High-Speed Broadband project also described. He worked from home, consumed entertainment at home, and demanded high performance from his technology. His domestic media ecology was correspondingly complex. He assembled his technology via online purchases, and, to make these purchases, he was frequently thrown back on his own technical skills and judgments, as he was when faced with assembling and then maintaining the devices. These complex ecologies demanded an expert, and commonly that expert was a DIY householder.

Once a device entered the home, digital housekeepers positioned it in the existing ecology of the home, which often involved getting the new device to "speak" to the network of existing ones. New technologies were not only accommodated by the householders, but they also needed to be accommodated by preexisting technologies. Depending on the technical affordances for plug-and-play devices, the degree or possibility of configurability, or the expertise available within the home, new devices might be selected based on perceptions of ease of establishing interoperability. In Anne and Michael's home (High-Speed Broadband 2004–2010), recent technology purchases had all been Apple products because they perceived the interoperability between Apple devices to be seamless. This was obviously not always the case, as they discovered when they realized that switching from a Mac to a MacBook meant their music no longer streamed to their TV without some reconfiguration, a job that Michael struggled to find time to attend to. Similarly, in Justine and Craig's family home (High-Speed Broadband 2004–2010), the introduction of a new router disrupted the preexisting family-media ecology, enabling their teenage son Dylan to access the internet through the night: "Justine used to be able to switch his internet off from our bedroom when she went to bed, which we haven't got the facility to do just at the moment . . . I've got to reinstall some stuff" (Craig, High-Speed Broadband 2004–2010). In each case, the labor required to get the devices to "talk" to one another, and to reestablish rhythms of family media use, was significant.

Digital housekeeping was especially laborious and frustrating when the interoperability of devices in the networked ecology was opaque and expertise fell short. For instance, Diane and Scott spent a considerable amount of time investigating why their wireless doorbell was ringing every time the living room lights were switched on using the light's wireless remote:

> When you turn the lights on the doorbell rings. The light is on remote control. It only happens after a blackout or power surge. We disconnected it for a month but it still didn't work. Disconnected it again and now it works. Sometimes the doorbell would ring at night. (Scott, High-Speed Broadband 2004–2010)

Such examples demonstrated that bringing new devices into the networked home brought about new complexities of interoperability, and the more devices there were in an ecology, the more complex their interactions became—a phenomenon of considerable interest as the Internet of Things emerged. Second, they demonstrated that expertise and digital housekeeping are not mutually constitutive. The ability to maintain systems of interoperability is linked to comprehension of how such systems function. With this in mind, below we consider what constituted digital expertise within the home, and what forms of digital expertise were available and valued within the home.

EXPERTISE

Expertise can be described as a techno-social construction that draws on existing social and technological dynamics, such as the political economy of technological production and materiality of digital systems through which the user is "configured" (Bassett et al. 2013). From our data, we identify three particular measures that households used to differentiate expertise: comprehension, knowledge transfer, and automation, each of which was performed in a variety of ways. For example, Christine (High-Speed Broadband 2004–2010) had a sound comprehension of the ecology and setup of her household's communal media system, setting up the computer connected to the TV to be controlled from an app on her iPad. To transfer this knowledge to others, she also set up the same app on her housemate Alexis's iPad. Peter (High-Speed Broadband 2004–2010) set up his MacBook Pro as a dual-boot system, with both Microsoft Windows and OS X (at the time, Apple's operating system for Macintosh computers) because his wife Stephanie was more familiar and comfortable with Microsoft Windows. There was also a distinction to be made between expertise in the use of preconfigured systems and expertise in constructing such systems. Ashley (High-Speed Broadband 2004–2010) was able to use complex media setups in tightly constrained ways, but she was not interested in understanding how the system functioned beyond the demands of her own use. In the outer suburban home of Antonio's family, there were similar variations in expertise. Antonio described himself as the "trendsetter" and "problem-solver" (High-Speed Broadband 2004–2010). He built his own PC. It had a top-range graphics card, two terabytes of data storage, an additional solid-state drive, a Blu-ray drive, and 16 GB of RAM for running the game Minecraft (all state of the art at the time). Antonio's younger brother Rodrigo had no such expertise or interest, but in his room was a similar computer, also built by Antonio.

Acquisition of expertise was motivated by a desire for comprehension as well as for pragmatic reasons. Diane and Scott, at the time living close to Antonio, were learning to comprehend systems. "It's hard," said Diane at our first visit (High-Speed Broadband 2004–2010). Diane used a computer at work but didn't feel she sufficiently understood how digital devices work. At home she tried to "fix things" in an effort to comprehend better. For instance, when we met, Diane had just learned how to configure her email accounts on her smartphone:

> I just learnt to set up email, and the difference between POP and IMAP addresses. I'm impressed because my brother couldn't work it out and he's in IT. Now I know the difference. POP address will only sync your inbox, IMAP will sync in and out. (Diane, High-Speed Broadband 2004–2010)

Scott similarly described himself as a novice. He slowly learned his way around things, and, in the process, surprised himself at what he was able to do. He described his knowledge of technology as "better than most people" (Scott, High-Speed Broadband 2004–2010). As Scott demonstrated, definitions of expert were relative to context.

The process of becoming expert involved a period of learning, whereby digital housekeeping skills were acquired and practiced in a deliberate manner. Deborah and Donald lived just outside a regional town on a wireless connection (High-Speed Broadband 2004–2010). Deborah and Donald's usage of their devices suggested more limited digital expertise compared to the participants already described. When showing their shared laptop to us during our first interview, Deborah conflated the Wi-Fi signal strength with the bandwidth. Deborah also told us she turned the laptop off after use so that spam emails could not "get through." However, since getting high-speed broadband, both Deborah and Donald had spent more time getting to understand the digital technologies already in their home. Specifically, Donald had become much more involved in the web-design process and had learned to manage his business's website himself, applying skills developed in the domestic environment to the professional workplace. He was creating all content for the website, learning basic web development and photo editing skills in the process. With this in mind, Deborah described Donald's acquisition of digital knowledge, and his shift in their social circles from novice to expert:

> He has learnt so much in the past seven months. Learnt so much! Now he's functioning on his own which is huge for him. He has never owned a computer in his life let alone had a website and done emails. And some of his friends would laugh, he had no interest. Now it is what he does, and now he sources work and controls his business. (Deborah, High-Speed Broadband 2004–2010)

Deborah and Donald, and Diane and Scott, indicated they were on a massive learning curve toward becoming competent, if not expert. At that stage, they could identify gaps in knowledge, and ask questions or take steps to address those gaps, but they were not yet in the position to transfer knowledge.

Being expert involves the ability to transfer knowledge. As Diane showed, there was a considerable difference between acquiring knowledge and being able to transfer it to another: "I need time for it to register. If Scott asks me a question, I can't explain it. I just know it works that way, and that's no good" (Diane,

High-Speed Broadband 2004–2010). Her understanding was not yet automated: "I can't do everything at once because I don't understand it enough. I'm just doing baby steps" (Diane, High-Speed Broadband 2004–2010).

Expertise is demonstrated in the automation of practice. The repetition of actions to the point of flow is indicative of expertise. Emily, as the "download queen" of her household, was more familiar with the processes of locating and downloading digital content than her siblings. Automation is evident when digital skills are sufficiently embedded to enable them to be exercised while multitasking. Antonio's younger brother Rodrigo often talked to his friends on Skype through his iPad while playing a massive multiplayer online game on his computer. The majority of participants identified rhythms of multitasking as indicative of their competences: Angela watched TV while searching for articles for her university essays on her iPad; Riley and Ashley browsed websites, checked emails, read blogs, and worked on their laptops on the couch while also watching TV (High-Speed Broadband 2004–2010).

Households where participants were in the process of becoming expert typically separated the locations of digital technologies, creating spaces of attention. Diane and Scott's PC was located in an upstairs study as a purposeful separation from leisure spaces:

> We made one of the rooms upstairs a study. I'd rather it all in one area. For me, computers are work. If I'm on the computer, it's work. I suppose I don't really see sitting on a computer as a fun thing it's more to get something done. Scott probably does it more as a hobby. He likes to look up things, whereas for me it's banking, or my dad's paperwork. I'm on a computer all day so the last thing I want to do is come back home and have to sit on a computer... I find I get caught on there. Before I know it, four

hours have gone and there's a quarter of my weekend. (Diane, High-Speed Broadband 2004–2010)

HOUSEKEEPERS AND THE DISTRIBUTION OF EXPERTISE IN THE HOME

A frequent argument in feminist literature highlights ongoing discrepancies in the distribution of domestic labor. Women perform a greater proportion of domestic labor, regardless of their financial contribution to the household (Hahn and Wilkins 2014). Forms of domestic labor are part of one's enrollment in the project of maintaining the household by contributing in ways that are visible, and visibility is grounded in threshold levels—the subjective point at which a person is stimulated to perform a task. Variations in threshold levels contribute to allocation of domestic labor and perceptions of expertise within those tasks:

> This pattern creates self-reinforcement, and the individual(s) with the lowest threshold will perform a given task even at low stimulus levels, until he/she becomes specialist for that task. Thus, when we apply this theory to human domestic labour, it suggests that my partner may begin doing the laundry because he has a lower threshold for piles of dirty laundry, but through repetition, he becomes "expert" at laundry. Ultimately, he and I will come to see the task as "his" and a self-organising system of domestic labor is created, reproduced, and maintained in everyday practice. (Alberts et al. 2011, 27)

Expertise is, in this way, also subjectively distributed in the home. Once it becomes distributed and habituated, it is difficult to renegotiate (Alberts et al. 2011, 32), and so it is with digital housekeeping. In their initial situating of digital housekeeping, Tolmie et al. "prefer to suspend the broad concerns with gender that occupy mainstream social scientists, and instead seek to inspect the particular demands of digital housekeeping from the perspective of *household members*" (2007, 333, emphasis in original), though they do not dispute the role of gender per se. Yet, as our analysis shows, it is problematic to situate digital housework outside of gender concerns when considering how housework is premised on particular forms of gender-inscribed expertise.

If we consider expertise as (1) comprehending systems, (2) the ability to transfer knowledge, and (3) automation of practice, then it is possible to make some judgments on where or, rather, with whom expertise is associated within each of the participating homes. A significant proportion of participants in our research who categorized themselves as "expert" were male. A significant proportion of female participants described themselves as being digitally literate and competent users of digital technologies but were "disinterested," or resigned to a male household member's expertise when it came to digital technology–oriented decision making or management. Often, it seems, expertise functioned as a proxy for interest. For example, Riley, an accountant, spent hours researching internet forums, sale websites, and consumer choice websites in order to make decisions on device purchases for his and Ashley's home:

RILEY: Ashley's not really interested.
ASHLEY: I just want things to work. I'm not interested in doing a whole lot of research on them to be honest.
RILEY: The server was a big deal. A lot of hours in that one. I wanted to make sure I got the right one.

ASHLEY: Other than my own phone I would say Riley chooses most of it. (Ashley and Riley, High-Speed Broadband 2004–2010)

Ashley and Riley had clear ideas about who would do the research into new technologies, as determined by their individual levels of interest. While it was clear that Ashley had interest in what technologies could do and was willing to research devices for her own use, she left the researching of home devices to Riley because she perceived it to be something he enjoyed doing more than her. She described his digital housekeeping as an interest rather than a skill.

As one would expect, access to interested expertise reduced motivation to become expert oneself. Because there were people in her household who knew how to download movies and TV shows better than her, Adele had little incentive to attempt this herself:

I don't know how to download. Tom does all that. People will come in and say, oh, what new movies have you got. Michelle will come in and ask for something on USB and she will watch in her room and come back or we will come in and say, Emily, can you download something? (Adele, High-Speed Broadband 2004–2010)

While Adele could locate and download content herself, it required more effort. It was, quite simply, easier and more convenient for her to ask Tom or Emily.

The expectation of expertise in the role of digital housekeeping had implications for other competing household pressures, and for the way those competing pressures were distributed among household members. Digital housekeeping took the place of attending to other domestic tasks for the digital housekeeper, which then had to be compensated for by other household members, also impacting the way those in the household created meaningful routines for themselves in the home.

For many of our expert participants, digital housekeeping was a component of their identity in and outside of the home. Digital housework, like other forms of domestic labor (Johnson and Lloyd 2004), contributes to identity and self-worth. It also affords visibility of the digital housekeeper's enrollment in the project of maintaining the household. Jeremy, as a university lecturer in network design and security in an engineering faculty, created complicated setups in his home in order to demonstrate them to his students:

> Internet access is important for me because that's my job. I teach students how to build internet devices, and build the internet and how to design it. I've got a server there. There's a lot of things I run on that box, which I don't need to run. If I wasn't teaching... I probably wouldn't have the full level of complexity that I have here. I often use the server here as a demonstrator in class, I log in to my home system to show people how things are set up in a working environment... I run a web server here, which hosts half-a-dozen websites, but the only people that access those sites are me. (Jeremy, High-Speed Broadband 2004–2010)

As a self-described "on-the-edge technology person," Antonio spent most of his disposable income on new devices. Riley and Ashley both called themselves "early adopters" in their social circles, although except for her mobile phone Ashley left all tech-purchase decision making to Riley. Participants adopted a discourse of "choice" in describing digital housekeeping that imitated arguments of women's domestic labor being a choice. Such arguments assert problematic power differentials:

> When involvement in housework is constructed as a choice, any disproportionate contributions can be defied as voluntary, the labour of others may not be demanded and attempts to change the behaviours of others cannot be legitimate. Even complaining is inappropriate because it misrepresents one's actions as somehow forced by others. (Natalier 2003, 266)

There are two competing issues here. The first is that labeling digital housekeeping as a choice based on interest undermines the centrality of the labor required to maintain the household. The second issue is that often the extreme degrees of complexity requiring additional labor *is* by choice. A proportion of the "work" of the digital expert in the home is based on interest rather than on efficiency or need. Often these interests create the need for more work within the household. Rather than contributing to the running of the household, this labor actually holds the power to disrupt it, as this quote from Jeremy illustrates:

> It's a complicated setup, so it is like a business. I have to spend a night a week to maintain it. To update it. Sometimes I will announce there will not be internet or television recordings for an hour. It depends, I don't do it at a convenient time. The worst is when I claim it will only be off for half an hour. It is usually only one of the internet or the media that is broken, so they can use the other while it's out. (Jeremy, High-Speed Broadband 2004–2010)

In hosting his complicated setup, Jeremy occasionally had to reset the entire household system to fix a bug. One of the few interjections by his wife, Amy, into our discussion was to tell us that she had few expectations of Jeremy's participation in broader housework, except that he keeps the technology working. A similar scenario played out in Malcolm and Nysha's home (High-Speed Broadband 2004–2010). Though Nysha was herself a technology expert, Malcolm was the driving force behind setting up the systems in their home. He enjoyed "mucking around with it as much as anything" (Malcolm, High-Speed Broadband 2004–2010).

Being able to "muck around" extended beyond processes of knowing or being expert. Several participants indicated that playing around with technologies in their home was a means of championing innovation, acquiring technologies that are not widely adopted in perceived support of technological advancement at the cost of efficiency or ease. It was also apparent through the data that those with expertise had the most agency in making decisions related to technology in the household (though there were of course exceptions), and that such decisions were not always to the advantage of the household. For instance, Nicholas was very excited to show us a Mesh Potato, a low-cost telephone and internet device, produced within an open-source philosophy, that enabled the user to make voice-over-internet-protocol (VoIP) calls through an old analog handset (High-Speed Broadband 2004–2010). His wife, Marwa, had no choice in the setup of the home phone system. She regularly tried to use the Mesh Potato to call her parents, who lived in Northern Victoria, a full day's drive away. The Mesh Potato had a tendency to crash, so Marwa often could not place or receive calls. Therefore, she often used her mobile phone or Skype to call her parents. Whether or not technology "works" was judged differently by experts and non-experts.

LABOR AND CONDITIONS OF EXPERTISE

Practices of digital housekeeping necessitated forms of immaterial labor. Many participants identified tasks required in their homes that they were hoping to "get round to doing." Craig liked to hack or tinker with systems. He had "a list" of things to do, such as connect the TV in the kitchen to the Wi-Fi router, reconfigure hard drives after getting a virus on them some months back, and reinstall software that would allow him and his wife, Janet, to curtail their teenage son Dylan's late-night internet use (High-Speed Broadband 2004–2010). Similarly, Jeremy described the amount of work it took to keep his home system running amid the pressures of everyday life:

> In here we have a computer connected to the TV, which is essentially the media centre for this room. It can stream videos from the central server, that's also the TV recording box, so it's essentially a TiVo box, home built, that stores and plays back all our recorded content. It is half broken, it records but doesn't play back at the moment because I haven't had time to fix it . . . When it was working better, before I broke it, this is where we sat and watched stuff. When you have a complicated house there are always things that are temporarily broken. (Jeremy, High-Speed Broadband 2004–2010)

Digital housekeeping was an ongoing process performed amid other competing pressures. Like other forms of domestic labor, digital housekeeping was cyclical. However, the material flows of technoculture meant that expertise was hard won and fleetingly held. Expertise was a constant dynamic process: "I think technology is moving far too quickly for me to understand it. I'm trying to grab hold of what I understand now and understand it properly so it will make sense later" (Diane, High-Speed Broadband 2004–2010). The possibility of becoming expert was constrained by the practicality of attaining a level that is incrementally mounting. Furthermore, expertise was conditional on access to ongoing opportunities.

The home is a site of considerable immaterial labor. To the already long list of tasks performed in the home it is necessary to add those associated with digital technologies. The home is not only a place of care, food production, education, laundering, and so forth; it is also embedded within social and technological constructions of expertise. The expertise required to build, maintain, and use the networked home media ecology is significant, though valued and acquired by members of households unevenly.

In our research in digital domesticity, we have found that, like other forms of domestic labor, digital housekeeping is not evenly distributed across all members of the household but clusters unevenly in particular individuals. We have also found that digital housekeeping, like its traditional domestic counterpart, is not evenly distributed across both genders, but, unlike traditional housekeeping, is more likely to be performed by men. The role of personal interest in this uneven distribution across individuals and genders has been found to be important in the eyes of our participants. Personal interest is associated with the acquisition of experience and, through experience, expertise, which spirals into a self-perpetuating cycle as others defer the work of digital housekeeping to the expert, thus building further experience and expertise. Personal interest and expertise are also self-perpetuating insomuch as the interested expert is inclined to construct complex systems as an expression of this personal interest, rather than as a pragmatic solution to a problem. The complex system is then inclined to remain in the domain of the interested expert, rather than being a resource and a responsibility shared by all. Extending this discussion of digital housekeeping and the distribution of expertise in maintaining a household ecology of media infrastructure, networks, data, and content, in Chapter 5 we examine negotiations that take place around household media use, with emphasis on the temporal rhythms at play in the domestic media ecology.

[5]

NEGOTIATIONS

In a scene that often played out in Australian suburbs in the early 2000s, Andy, age fifteen, was meant to be concentrating on his homework but was actually on MSN[1] chatting with friends (Connected Homes 2004–2010). When he heard his mother's steps in the passage, he deftly minimized the MSN window. As she walked by, she peered in to make sure that he was doing his homework—he was, assiduously. As the sounds of her footsteps receded into the kitchen, Andy reopened MSN and messaged his friend informing him that he would definitely attend the party on Saturday night. But, later that night, Andy was disappointed to find that his parents would not allow him to go. The party would be in a distant suburb, and, from his parents' perspective, the party had been arranged through a worryingly abstract set of relationships transacted on the internet. Despite a week of discussions, arguments, and any number of strategies to redefine the term "party," Andy failed to convince his exhausted but resolute parents otherwise. This incident seemed to compound the parents' growing anxiety about their son's persistent internet use, the time he spent in chatrooms, and the sorts of relationships he

---

1. The term "MSN" or "Microsoft Network" was used to brand many services over the years but at the time was synonymous with Microsoft Messenger, an instant messaging client later rebranded Windows Live.

Digital Domesticity. Jenny Kennedy, Michael Arnold, Martin Gibbs, Bjorn Nansen, and Rowan Wilken, Oxford University Press (2020). © Oxford University Press.
DOI: 10.1093/oso/9780190905781.001.0001

developed "there," many of which remained uncomfortably intangible and indistinct to Andy's parents.

PARENTING AND MEDIA ECOLOGIES

Andy's parents, Annie and Robert, were both in their late forties and also had a nine-year-old daughter named Cathy. Annie consciously steered her children's technology use toward positive, mainly educational, ends. As a linguist and employee of a public educational research institute, she avidly sought out articles, books, and computer programs that would further her children's education and had bought a computer program to help Cathy with her grammar. Looking for something for Andy, in her "domestic probe" diary (see Chapter 2), she noted: "I visited an educational products site in the US—one of the many that appeal to the sometimes insecure 'am I doing enough for my child' syndrome. Decided yep I am!" (Diary, Connected Homes 2004–2010).

Annie's search for computer educational programs had intensified as her concerns about her children's use of technology had grown. In the year between our first two visits to the home, Andy's internet use had become "a major concern" for Annie, and this concern and the conflicts surrounding it dominated much of our discussions with the family about parenting and the children's use of media. She was anxious about the amount of time he spent online and, in particular, his MSN chatting. We observed a palpable tension between mother and son when the subject of Andy's use of the internet came up, although this was diffused by Andy's joking manner. For instance, Andy teased his mother by referring to the way he strategically minimized the MSN window when Annie entered the study or walked past. "Now that we've got two doors opened up, it's much harder to hide," Andy quipped, "but magic Alt-F4 works every time!" (Connected Homes 2004–2010).

We read in Andy's probe diary unequivocal enthusiasm for chatting: "I love MSN"; "I hate it when the internet is down." "Being the person I am," said one diary entry, "on the days before November 4th, I used the computer for research for about 1 hour and the rest of the time I was on MSN." Accordingly, one of Annie's principal concerns was the way Andy "switches between MSN and homework constantly and that's when I start to—"

"It's very difficult," Robert interrupted her, "for him to get any concentration momentum. It's a real problem." Andy, however, was defiant:

> But you need them [MSN friends] there because then they can contact you if they need to because . . . like we're doing like a group . . . um . . . work thing at the moment so I had to send my friends a copy of something, so we could fix it up and finish it off and get it handed in tomorrow. (Andy, Connected Homes 2004–2010)

Both parents appeared skeptical. "Yeah," says Robert wearily. Annie and Robert also questioned the quality of the interactions that took place in chatrooms. As Annie noted: "There doesn't seem to be much language exchange—LOL, WDF, OMG, short yes/no exchanges and expletives." But it was not just the time spent online, the distraction from homework, and the "impoverished" online linguistic conventions that bothered Annie and Robert: They were principally troubled about the potentially pernicious relationships Andy might develop online. Annie felt disturbed by what she saw of Andy's MSN message content and heard of his telephone conversations: "I'm not happy with this aggression that seems to mark the boys' relationships with each other." She was particularly vexed by the emergence of bullying behaviors. When Andy admitted nonchalantly that he preferred internet to texting because "you can't block on a mobile, but you can block on the internet" (effectively excluding certain individuals from the social group), Annie became upset. As it happened, one boy in Andy's grade tried to join his social group but was continually facing rejection. "They snub him at school and then on MSN he keeps having to change his 'name' because they keep blocking it," Annie told us, shaking her head.

Animated by an article on cyberbullying that she downloaded at work, Annie elaborated in her probe diary:

> [The article] covered issues that I have been concerned about and witnessed here at home with Alex and his MSN chats. There is one student in the class whose social skills are in need of attention. Unfortunately, he is a target for being ostracised. MSN makes this possible to continue beyond school. He is blocked from chats. I have witnessed some of this and raised it with Andy. As a parent, I thought it appropriate to raise it with the school, which has acted on my concerns. I agree with the researcher in the article who states, "Kids' online and offline lives are so intermingled now—they chat online at all times of the day and night, whereas with traditional bullying home used to be a safe place." (Annie, Diary, Connected Homes 2004–2010)

On an optimistic note, Annie hoped that a positive outcome of "this potentially destructive and insidious [technological] development" would be the opportunity it afforded to bring bullying into the open, "on the screen and in your own study." Andy's parents feel disconcerted not only by the social exclusion that MSN facilitated, but also by the social connections it enabled. Andy's father, Robert, noted that Andy "is bolder with girls on MSN than with girls in real life." Andy was acutely aware of, and immensely pleased with, the advantages of MSN when it comes to girls:

> You can't see each other on MSN, they can't run away when they see me. It's different, [they're] easier to talk to. It's less embarrassing, you don't have to deal with silences, and if you don't like what you've said you can just backspace. (Andy, Connected Homes 2004–2010)

Most of Andy's MSN "girlfriends" remained anonymous, although they sent photographs. "They could send me a photo of a supermodel and I wouldn't know the difference," Andy laughs. Eventually, however, some possibilities arose for Andy to meet one (or several) of these girls. Such an opportunity came to pass during our research, and we were able to witness the escalating tension between mother and son. Andy had been invited to a party by a girl who was a friend of a friend on MSN. Annie was nervous:

> A dilemma, and a new one at that—parties have not been on the social agenda as yet. The girl lives in Mitcham. Who is she? Who'll be there? How old are they? Will there be adult supervision? I'm sounding like my parents! It's so far out of our community! I have no sense of what it might be like. So, no, we won't consent! (Annie, Connected Homes 2004–2010)

Annie explained to us how she and Robert were uncomfortable with Andy going to a party "on the other side of town" and that this led to "a major conflict" between son and parents. In the wake of this tension, another disagreement took place when Annie and Cathy came home after a weekend away to find that Andy's attention had temporarily shifted from MSN to a new videogame. Annie reports:

> Andy and I had another altercation. He has bought a new PlayStation 2 game and from the sounds of things has done a lot of playing while we were away. Tonight, he came home from school and was straight onto it. I told him to get off and do his homework first. I had to threaten withdrawal of privileges before he listened to me. I loathe these conflicts. And I loathe the effect of the games on his (aggressive) behaviour towards me. (Annie, Diary, Connected Homes 2004–2010)

Andy had a more sanguine view of his own technology use. For example, he considered his use of MSN rather modest compared to that of some of his friends who were "chronic users of MSN—it's like, 'Go get a life!'" Andy exclaimed. At times, even the parents were willing to put their concerns into broader perspective. "There are households," says Annie, "where adolescent boys are glued to the screen, and that would be a major concern to me. He's pretty sensible really." "Yeah, he's pretty reasonable," Robert agreed.

However, their perspective shifted again when they considered the broader dangers of internet chatrooms. Prompted by a televised debate on pedophiles' online access to children and hundreds of pornography-related arrests, Annie noted:

> Interesting concepts that are very relevant to me as parent: the notion that the internet is a public space and therefore the potential for abuse is as high as it is with the more conventional ones—parks for instance. The issue here is of course to educate children in understanding what some of the dangers might be

> when they're in those chatrooms. (Annie, Diary, Connected Homes 2004–2010)

Throughout all of our research we noted that parents struggled in their attempts to weigh up the risks and restrain their children's behaviors in an appropriate manner without unduly restricting or jeopardizing their children's growing independence. With some exasperation, Annie asked, "But how do I act on it? Where's the balance between what I see, which comes into the public space on the screen, and Andy's right to privacy?"

Four years later we revisited the family and found that Annie did not hold the same concerns for her daughter, Cathy, despite Cathy being three years younger than Andy was four years ago. Instead Annie argued that it was important that Cathy use new technologies (such as MSN) so that she could "fit in" socially. This view was echoed by other parents in the research. Annie's reversal in her position on the use of MSN could be linked to her increased exposure to MSN through her work. As MSN had been made ordinary for her, it has lost its exotic character, and, with it, her feelings of anxiety that were associated with its strangeness had reversed. This had also coincided with a change in attitude on Andy's part; once a vital part of his existence, MSN was no longer exciting or indispensable. For Andy, MSN was no longer essential to his social being—essential in giving him access to his social world, essential as a mark or symbol of that membership, nor essential as an assertion of the nascent identity of an independent adult. MSN had retained some utilitarian value but was no longer invested with the glamour and the connotations it once possessed for Andy. It had become passé. Not at all passé, though, was Andy's new passion for learning to drive, and Annie's anxiety about Andy's internet use had been displaced by an anxiety about Andy behind the wheel.

PARENTAL STRATEGIES IN DOMESTIC MEDIA ECOLOGIES

Over the past two decades we visited many families like Annie and Robert's and talked with them about their experiences and their parenting strategies in the face of new technologies (Wired Homes 2002–2006; Connected Homes 2004–2010; High-Speed Broadband 2011–2017). In this section, we identify the major strategies and stances and contextualize their nuances and subtleties vis-à-vis the particulars of the family relationships. We also place our findings in the context of the literature on families and technology use, relating the particularities of the vignettes to observations derived from quantitative and larger-scale studies.

To suggest that parents employed particular strategies or stances is in itself problematic. In the first place, the parents often expressed individual positions on many issues rather than presenting a united front. They also tended not to make generalizations about technologies as a whole, or even about particular devices such as television, or the internet, or games, or even about a particular television show, a particular site, or a particular mode of device use. The approach to digital technologies was often tentative and uncertain and, in our experience, invariably nuanced. Parents revealed a great degree of thoughtfulness but very often remained inconclusive and wavering on their children's use of computers and social media. The highly critical and discriminating stance of parents toward children's technology use tended to be associated with ongoing argument, reassessment, negotiation, and tentative action. Accordingly, some parents preferred not to make general overarching rules based on unchanging principles as much as they performed context-specific "executive decrees," which settled the matter at hand until the next round of assessment and argument. Other parents did establish rules (no television in the bedroom, no internet until homework is done), albeit flexible and negotiable rules, but, either way, particular device use and rules were centers of ongoing negotiation and contestation, through time, and dependent on the contingencies of the day. Interestingly, none of our participating families automated the enforcement of rules through deploying internet filtering and similar software to control and monitor their children's access to online content (except one who would manually switch off the Wi-Fi in the home to curb their teenage son's late-night gaming [High-Speed Broadband 2011–2017]).

Contingent and negotiable though they might be in practice, many parents in our earlier research expressed a lack of clearly stated or enforced rules on their children's device use. Yet, in the early part of the twenty-first century, as the internet was joining the domestic media ecology, all of our participating parents expressed some type of concern about their children's internet use. This was reflected in survey data, with almost all Australian parents at this time (92 percent) expressing issues of concern in relation to the internet, while simultaneously almost all (99 percent) regarded it as being of some value (NetRatings Australia 2005). Since this time, tensions between parents' perceived risks and benefits of digital media and internet use have persisted (e.g., Livingstone 2009; Nansen et al. 2012; Ofcom 2017). These hopes and fears have, of course, migrated and widened to accompany the rapid growth in the number of mobile and internet-connected devices in family homes (e.g., Ofcom 2017; Marsh et al. 2018), as well as the increasingly younger age of children using mobile and internet-connected media (e.g., Plowman et al. 2010; Clark 2013; Green et al. 2014).

Alongside these concerns, research highlighted the changing patterns in the ways parents regulate their children's media use. Referred to as parental mediation, this was originally developed in reference to the medium of television and the types of rules and restrictions imposed by parents on the routines and content of children's viewing (e.g., Bybee et al. 1982; Austin 1993; Pasquier 2001). Studies have since extended the focus on television to include the moderation of newer media, such as computers, internet, and videogames in the home (e.g., Nikken and Jansz 2006; Livingstone and Helsper 2008; Valcke et al. 2010), as well as children's use of mobile devices (Nansen and Jayemanne 2016; Nevski and Siibak 2016a). Different styles of parental mediation have been identified and categorized in terms of restrictive mediation, active mediation, and co-viewing or co-playing (e.g., Nikken and Jansz 2006); that is, restricting media use, talking about media use, and viewing or sharing use, respectively. These different styles have been located along a spectrum from more authoritarian to more autonomous approaches (Green, Holloway, and Quin 2004) to analyze how parents perceive, place, and manage media in the home.

Research shows that styles of parental mediation are informed by factors such as children's age and type of media, parents' views about the positive or negative effects of media on children, and the kind of media environment and routines families have established (Downes 2002; Valcke et al. 2010; Ofcom 2017). For example, over 90 percent of those surveyed in Downes's 2002 Australian study established family rules governing computer use, couched in terms of priorities granted to certain activities over others, limitations on time and content, and rules designed to protect the technology itself. International studies of internet use, television, and computer gaming also found that rules of some kind were almost universal and most commonly took the form of time-based restrictions, content-based restrictions, and a requirement for parental participation (Nikken 2003; Koolstra and Lucassen 2004).

Yet parental decision making is complicated by research highlighting a wider set of arrangements and influences, including household media ecologies, touchscreen affordances, and sibling relationships (e.g., Nansen and Jayemanne 2016; Nevski and Siibak 2016b). In addition to the temporality of media negotiations, studies of the spatial dynamics of family media interaction reveal situations in which children actively push back and attempt to redefine the rules and uses of media in the home (e.g., Aarsand and Aronsson 2009; Pasquier 2001). Such negotiations challenge assumptions that locate control within a disciplinary model in which parents impose media rules and children dutifully follow them. As our informants told us, the internet, television, computer games, and other forms of devices and digital media are ever-changing, as are the children, the parents, and the world, and this ongoing transition sets up circumstances for the moving feast of internet-focused critique, negotiation, argument, and decree. Digital devices provide an important resource for parents and children to play out developing skills and experience in negotiation, argument, debate, compromise, winning and losing, taking responsibility, and conflict management.

Accordingly, negotiations around rules associated with device use—particularly chat, email, and gaming—become anchor points for our participating parents to declare and fix their stand on important principles and issues, such as the work ethic, good health, violence and exploitation, bullying, feminism, and their own authority as parents. This stand was reasserted day by day with each new television show, computer game, or website. Similarly, device use anchored our participating children's expressed position and principles in regard to their own sense of maturity and independence, and to their work ethic, health and safety, and values and morals. Technology thus provided a focus for what a parent is and does and should be, and what a child is and does and should be, and this focus ran thematically through the negotiations, in a transient and particularized way. The point is that rules and negotiations do not just circulate around the unchanging desirable and undesirable qualities of technology; rather, they circulate around the changing and particular qualities of particular technology, and the desirable and undesirable qualities of parents and children.

Opportunities were identified for positive experience with technology—particularly in relation to education, and as an occasion for family time together. Our participants across all projects thus joined the overwhelming majority of parents in the view that, from the get-go, the internet could be of benefit to children's study, and our participating children routinely used Google, Wikipedia, and school-directed internet resources for these purposes. Our participating families also valued technology (overwhelmingly television, video, and gaming) for the focal point it could provide for social participation within the family. Contrary to generalizations to the effect that television watching is, by and large, a nonsocial or even antisocial activity, in the early years of this century our participants provided examples of viewing television, and then DVDs and YouTube, as valued socially productive activities. They reported that movies and television shows and, less frequently, internet sites and music were set up to provide a shared experience that stimulated conversation and other communal activities, and provided a source of solidarity and commonality in shared responses to storylines and characters. In addition to identifying these social goods, some of the older literature goes further and reports that television is used to block hostile interactions among family members, particularly in crowded households (Rosenblatt and Cunningham 1976), that it smooths conversational flow by filling in what might otherwise be awkward silences, and that it gives a pretext for couples and parents and children to sit in close proximity and display physical affection

(Morley 1986). While we recognized the multiple uses and multiple contexts of media, and our participants did value some media, the positives of streamed and broadcast media were seen in our research as the exception and not the default position; our participants indicated that these positive uses need to be actively sought out, encouraged, and exploited. They did not arise without active intervention and were as much the construction of the family as the construction of the television producer or streaming service.

Our participating parents were at best suspicious and at worst highly critical of media content. In their critique of free-to-air television content, they expressed a sophisticated assessment of narrative, plot and character development, cinematography, scripting, and the like. Many of our participating parents were uncomfortable with commercial advertising and the promotion of values associated with consumer culture; with the never-ending narrative structure of soap operas; with the valorization of inappropriate role models (particularly for girls); and with storylines and characters that are generalized as being degrading at worst and banal and insipid at best. As was been reported many times in other studies (e.g., Strasburger and Donnerstein 1999), news and current affairs programs representing violence and death were of great concern. That children witnessed tens of thousands of acts of violence each year on television was a claim most parents were familiar with and concerned about. Regardless of nuanced particularization of benefit, our participating parents appeared to agree with the generalization that television has undesirable effects on children, youth, and society in general (Fabes, Wilson, and Christopher 1989). However, this appeared not to be the view of the children, who either enjoyed watching television or, if not actually enthused, were at least more sanguine than their parents about any failure to maximize every opportunity to invest in life's experience. In Australia, parents and children also differed on other issues: parents' estimates of children's television viewing times were typically significantly less than the children's own estimates; fewer children shared parental concerns about the internet; levels of intervention reported by parents were usually higher than that reported by children; and the clear distinction that parents were inclined to make between work and play on the internet was not nearly so clear to their children (see Downes 2002; Koolstra and Lucassen 2004; NetRatings 2005; Australian Communications and Media Authority [ACMA] 2007).

It was clear that parents were dissatisfied with the quality of design and manufacture of free-to-air products and were prepared to look for alternatives, and, in the early years of our fieldwork, the alternative took the form of DVD selection. By selecting the DVD at the video store, and by thus controlling the material presence of the particular videos or DVDs in the house, parents were controlling physical access to content and asserting quality control. Where material presence is not an issue, parents are thrown back on the much more difficult task of using immaterial rules, exhortations, and admonishments.

Of course, all of this became still more difficult in more recent years with the redundancy of video and disc and the ready availability of streamed content, which provided undifferentiated access to billions of videos that, in many cases, had been through no quality-control measures whatever. Compounding the difficulties parents reported in selecting suitable "sharable" free-to-air content, let alone streamed content, was the shift in domestic media hardware, from large single screens shared by all in the living room or dedicated media room, to multiple screens "owned" by individual family members, viewed by individual family members, with content selected by individual family members (Figure 5.1). The proliferation of these devices over the past ten years eviscerated the

Figure 5.1. Children on personal or shared mobile devices engaging with individual games and streaming content (High-Speed Broadband 2011–2017)

negotiation of quality-controlled content selection and the shared experience of that content.

However, of far greater concern to parents than poor-quality content was the use of technology by unknown "others" to communicate directly with children without the opportunity for parental intervention. As the technology made its way into Australian homes, most parents accepted that their children might use the internet to chat in private, raising the specter of pedophile rings and online sexual predation by older children or adults, which has received sensational public attention. Most parents were mindful of this risk, whether real or not. Yet parents were not only concerned about the harm that might flow from technology to their children, they were also concerned about harm that their children might initiate through digital technologies. As we saw in the case of Andy and his parents (Connected Homes 2004–2010), sending out indiscriminate invitations to parties, bullying other children through SMS and email, and, in other cases, seeking information on drug taking or bomb making, and downloading pornography, were the kinds of activities that received public attention and were the kinds of activities that parents were afraid that their children might initiate. Of these sources of moral panic, access to pornography had historically been the issue of greatest concern to Australian parents, followed by communication with strangers (NetRatings Australia 2005). A concern with sexual content and sexual contact had also been apparent in international research (Livingstone 2003, 2009). Over time, adding to these content and contact risks were a range of more everyday risks associated with increased use of social and mobile media by ever-younger children, including harassment and bullying, along with exposure to sophisticated commercial and advertising content, and peer pressure associated with sharing visual content and body image (e.g., Ofcom 2017).

Today, technology use is seen in the context of children's growing maturity and relative independence. Domestic technologies have become instrumental and iconic indicators of ages and stages of childhood development—in the sense that being permitted to sleep over at a friend's house, having one's own door key, and being allowed to drive the family car are some of the more traditional markers of stages of maturity. Having one's own mobile phone, social media account, computer. and iPad is dependent not just on affordability and desire, but on parental attitudes to age-appropriate experiences for children, and on children's assertions of growing autonomy, materialized in their demands for ownership and use of technology.

The parents in our studies wished that family politics in respect to technology were simpler. If only they could cut through the exhausting rounds of critique, negotiation, argument, and decree that accompanied each variation in domestic media presence and with each new stage in their children's independence; if only there was an alternative to wave after wave of admonishment, reasoning, cajoling, raised voices, bargaining, and dire warnings. In this context, a strategy that was employed, one way or another, by all the parents in our research was to translate their ethereal will into the brute force of material reality. That is, an immaterial cultural, moral, or ideological position was translated into action, not through ephemeral reasoning and argument, but through the material arrangements in the house, through changing what politicians sometimes call "the facts on the ground." Just as speed bumps literally made concrete a moral imperative to drive more slowly (Latour 1992), and low-slung bridges made concrete a desire that certain people should visit the beach and others should not (Winner 1986), and numerical lathes made concrete a policy to exclude unionized labor (Noble 1979), so the desires of parents to manage technology were translated from ideational reasoning and symbolic foot stamping into hard facts on the ground.

In the early days of these new technologies, we saw that certain material provisions were made (DVDs and videotapes, computers in public areas, educational software) and certain material deprivations were put in place (certain DVDs were denied, an email account was denied, an internet connection was denied, a computer in the bedroom was denied) (Wired Homes 2002–2006; Connected Homes 2004–2010). Of these material arrangements, perhaps the most common was the translation of incidental observation into purposeful surveillance through the strategic location of the computer in a communal area—alas, a strategy that, in an age of personal smart devices, was much more difficult. In those days, almost half of Australian children accessed the internet in a study or home office (in a population skewed toward households with high income) and a further 25 percent from the living room. Unlike today, very few Australian children accessed the internet from their bedroom, whereas in recent years portable, personal media technologies have proliferated.

As the Australian Bureau of Statistics (ABS 2018) reports:

> The mean number of devices used to access the internet at home per household has increased from 5.8 in 2014-15 to 6.2 in 2016-17. For households with children aged under 15 years, the mean number of devices used was 7.8, compared with 5.4 devices for households without children under 15. Nearly all (99%) households with children under 15 used a mobile or smart phone to access the internet.

In addition, children aged 15 to 17 years are the highest proportion of internet users (98 percent) (ABS 2018). Similar figures pertain in other technologically developed countries.

The shift within the home from large, immobile, centrally located screens and internet connections to decentralized, mobile, individual internet connections altered domestic media ecologies in important ways. Physical control of access to content and to viewing devices is much more difficult in today's home, and parents were thrown back on immaterial advice and rules, significantly reducing their capacity for command and control.

Today, American children aged 0 to 8 use screen media for an average of 2 hours 19 minutes each day, and about 35 percent of this screen time is spent with a mobile device, compared to 4 percent in 2011 (Rasmussen 2017). That children spend "too long" in front of the screen is the generalization most often made by educators, psychologists, and physicians, and it is based on the view that the opportunity costs associated with technology use are too high. Consistent with this, our participating parents felt that time spent in front of a screen was a lost opportunity for outdoor play, talking with family, reading books, engaging with a hobby, doing homework—all desirable activities seen as at risk of being squeezed out of the day by the seductions of digital media. From the parental perspective, portable, personalized digital media and communication had no boundaries—no beginning and no end—and were seen as a bottomless black hole sucking out the potential for their children to live full lives made up of rich and varied experiences. The infinite horizons of digital consumption in this new media ecology provided no inherent endpoint, and, in its absence, a beginning and an end had to be imposed from without, by the parent, commonly around several turning points: a set maximum time; an absence of a clash with dinner or any other "more important" activities; and as a "reward" for completion of a duty, such as homework or dishwashing. Once negotiated, parents attempted to conserve energy by translating the negotiated outcome on time limitations into an immutable and generalizable rule, by ensuring that screens were absent from eating areas, for example, or from bedrooms after bedtime, but we found that children were quick to identify contingencies and changes in context, and negotiations are reopened time and time again.

Energizing these concerns was the question of control. All parents who participated in our research saw it as their difficult duty to control digital media to the extent that they impinged on the lives of their children and to control their children's exposure to digital media. And, as we have seen in our research, neither children nor digital media were particularly tractable in this new distributed, decentralized media ecology. These new media technologies were a focus of contested and ongoing negotiation, and these negotiations formed an important occasion for the ongoing negotiation of the parent–child relationship.

It may appear something of a paradox that while the dominant parental attitude was one of uneasiness and suspicion, their homes abounded in these technologies. Yet parents were concerned about many things in relation to their children. In a sense, everything is of concern to parents, and one concern is not neatly distinguished from another. Rather, our conversations with parents suggested that a good deal of parenting is a process, rather like dealing with a huge ball of tangled concerns, each thread of which is knotted up with a whole number of others. Certainly, parents were concerned about the banality of commercial television, sexism and racism on Reddit, bullying on Facebook, narcissism on Instagram, violence in videogames, pornography and other unpleasant surprises on websites, spam on email, strange people on chat sites, and so on. But they were also concerned about junk food, alcohol and drugs, their children's friendships, strangers, schools, sex, disrespectful language in music, supervised parties, unsupervised parties, and many other things that were not separated from domestic media ecologies. Media use was knotted up with food, school, and gender roles; messaging with friends, strangers, literacy, parties, and alcohol; web browsing with wasting time, sex, and prejudice; and so on (boyd 2014).

In our parents' accounts, sexism, violence, creativity, friendship, fun, and banality were not abstract notions that "stood behind" technology in another social or cultural realm (Latour 2005). They were materialized in the home in the life of the domestic media ecology. Symmetrically, being a parent was not a sociological abstraction that "stood behind" and informed the management of this ecology; rather, it inhered in the management of domestic media, just as parenting inhered in changing diapers, holding hands, feeding, and advising. Parenting is a flesh-and-blood performance that was to be found in the reflexive playing out of material relations among parents, children, diapers, roads, food, and digital media. It was contingent, contextualized, and improvised in conjunction with other actors. The technology, the content it delivers, and the good and ill in the world at large were all woven of the same thread; the threads were all knotted and tangled in a single sociotechnical ball that could only be picked at and rolled around and around. For the parents, it was all Sisyphus, and there is no Alexander in sight.

DOMESTIC RHYTHMS

Technologies act *in* time and *on* time. Control over time and negotiating the expenditure of time was a frequently occurring motif in our conversations with parents and children. Defining and then enacting bedtime, dinnertime, homework time, playtime, time to work, time to rest, time to sleep, and so on was far from simple in conditions where too much time was wasted here, too little time was available there, and time lurched between dragging and flying. Georg Simmel (2004), Paul Virilio (2006), and Judy Wajcman and Nigel Dodd (2016), among many others, observed the speeding up of time as technologies conspire with demands for personal and commercial efficiency to draw our attention away from the passage of the seasons turning into the passage of nanoseconds, and to insist on ever more coordination and integration as the rhythms of life are accelerated. This sense of urgency and impatience is palpable and pervasive enough in contemporary life to provoke pushback in the form of movements such as "slow food," "mindfulness," "technology-free zones," and the like. We identified the negotiated rhythms (Lefebvre 2004) of temporal dynamics at play in the domestic media ecology and describe these below through acoustic metaphors. We identify four forms of interlaced rhythms: (1) a polyphonic drone, (2) a polychromic dissonance, (3) an asynchronous consonance, and (4) orchestrating domesticity.

A Polyphonic Drone

Digital media research has observed that contemporary technologies provide resources for maintaining persistent availability, or a continual connected presence (Licoppe 2004). While our research similarly suggested that the process of integrating new media technologies into the home was associated more forms of availability and connectivity, and more entertainment and information options, constructing a milieu commonly experienced as "always connected" or "always on" (Baron 2008). This experience of always on was an ambient media presence that in acoustic terms is a *polyphonic drone* that constitutes the requisite background for all sorts of foregrounded activity in the domestic ecology. While normal, ordinary, everyday, ever-present, and always on, the noise produced by household media ecologies ironically receded from view, masked by its ordinariness and constant presence. The Instagram posts came in, along with Facebook notifications and status updates, music played quietly or loudly in the background, Messenger screens were always open, Siri listened for requests, mailboxes filled and flags were raised, YouTube videos were recommended, the smartphone vibrated in the pocket, the flat screen was on the wall, the laptop was on the table, an avatar danced as it waited for someone's attention, and casual games were running. In establishing an unbroken cacophony, the acoustics of the media drone constituted a solid wall of white noise and, in a sense, removed an awareness of the media and the noise they make. In so doing, it erased an awareness of the rhythms of time:

> You don't realise how much you use it [technology] . . . it's just there all the time; basically, I am always connected. (Maurice, Connected Homes 2004–2010)

> [I]t's just there as an online tool, whereas previously [. . .] it was almost you'd say, 'okay I am making the decision to go on to the internet and I'm going to spend some time on the internet' [. . .] whereas now it's like having an always-on connection. (Annika, Connected Homes 2004–2010)

Moreover, this ambient media drone was polyphonic, bringing together and compounding the media streams provided by television, radio, web, social networking, texting, messaging, video and music streaming, and telephoning, channeling a literally endless stream of voices, images, words, and sounds into the home. We refer to this as a polyphonic drone, which can be muted but cannot be switched off. This drone was seen by households to emanate from the nature of the media ecology itself, a technological imposition determined by the *telos* and character of the ecology. This phenomenon was most often noted in relation to screen media, particularly the television and, later, streamed video: "It's just endless. They [the children] can just watch the next show, and the next one, and the next one" (Mary, Connected Homes 2004–2010).

The drone of contemporary media ecologies drew an end to media forms that were discrete, had a beginning and an end, and were self-contained:

> The film is now only one nodule within a whole concatenation of options—playing a game, seeing how it was made, seeing

various versions of the director's cut with alternate endings, the out-takes, interviews—and so the film just becomes one other bit of merchandising . . . I see that as being a really significant shift. (Bob, Connected Homes 2004–2010)

Unsurprisingly, perhaps, the drone of household entertainment and interpersonal communication media was generally not perceived by participants in our research to be as invasive as the drone of work-related media. The ever-present possibility for knowledge work to be done (Gregg 2013), here and now, suggested the need that it should be done, here and now, constituting a form of tinnitus that, for many parents and adult householders, was hard to live with. Work that was materially (that is, digitally) present at all times translated into a rhythm that was flexible yet lacked the modularization of a beginning and an end. The drone was never-ending and ever-present, so one tuned in and out during the day, through the evening and on the weekends. Though one paused, the drone itself never finished:

> Some days I might have been on my machine in the office for most of the day. And I come home, and I'm so habituated to that space of the screen that I keep wanting to go [and use the laptop]. I have to say to myself, "no, hang on a minute, you've been working all day on this, leave it alone." But I feel a real need to still be online. And I'm beginning to really understand that notion of being wired, being on all the time. (Neil, Connected Homes 2004–2010)

Moreover, the capacity for speed in the contemporary communications environment created a demand for speed in an escalating positive feedback loop, where quick replies to the incoming stream established and locked in expectations of quick replies, compounded by a sense of urgency: "There used to be delay in communication in business, people used to accept taking time to get back to them. But now people expect quick answers" (Eileen, Connected Homes 2004–2010).

While these new conditions were in many ways a product of the ecology of contemporary digital domesticity, the appeal of this polyphonic drone was also framed by our participants in terms of their own agency and subjectivity, problematic though they saw it to be:

> [I]t's a problem with people at work, because there is an expectation that I am available. That's a really big problem. I've created it. They haven't. I've created ... What is it in me that makes me want to do that? Why can't I just switch off? (Yukiko, Connected Homes 2004–2010)

Nor were the effects of the drone confined just to adults and to work:

> Sana will go onto Club Penguin. And she will spend ages ... and I will say, "what are you getting out of this? All you are doing is spending time. It is like a prison sentence of sorts" ... They are designed to suck you in and make you spend more and more time, and you have to see that. (Georgie, Connected Homes 2004–2010)

In addition, parental concern pivoted on detrimental impacts for the child's body:

> [O]ccasionally at night, I have gone to bed and I've just been aware or sensed that the kids aren't asleep, and the next day he is all bleary-eyed ... this is the problem, I think, the one thing they really need to learn—the hardest thing—is to know how long they have been watching and to monitor themselves. (Sam, Connected Homes 2004–2010)

A Polychronic Dissonance

As well as affording expectations and compulsions to be "switched on," the polyphonic drone of the changing media ecology unsettled the temporal rhythms of activity that held sway in a differently mediated domestic past. A media ecology that is always switched on provides resources and their possibilities, as well as imperatives, to organize the temporal order of activity in novel ways. We refer to this new temporal order of activity and negotiation as *polychronic*, meaning that routines were not organized in a linear, sequential, and ordered manner; rather, the ecology offered the potential for ordering tasks in time so that they were frequently started, stopped, switched, and interspersed with each other in a way that was frequently dissonant. Distinct from the often strict, regulated, and linear time-task management of "monochronic" industrial labor (Thompson 1967), polychronic labor in contemporary media ecologies is characteristically nonlinear and nonsequential and is interpolated by technologies that are dissonant: they interrupt, distract, and demand immediate attention. In these ways, they might be regarded as both *polychronic* and *dissonant* (Hall 1959; Bluedorn et al. 1992; Daly 1996; Lee 1999; Lee and Liebenau 2002).

Polychronic dissonance was particularly evident in our studies among younger participants, such as Andy, who often overlapped or simultaneously conducted activities or interactions through multiple media devices and platforms:

> I talk to my friends and type sometimes, and then sometimes I have the music on when I do homework. And the small TV we have on sometimes when Mum is in there . . . It is usual that I do more than one thing at a time. (Andy, Connected Homes 2004–2010)

But these polychronic rhythms were less easily integrated by others. In these cases, the unpredictable nature of frequent interactions with media ecologies was viewed as an inconvenience and as destabilizing an ordered flow of everyday activities. This was particularly the case when considering the effects of synchronous media (where communication occurs in real time). While asynchronous elements of the media ecology (where communication may be delayed) asserted demands for attention in their own ways, the demands of synchronous media were particularly shrill and particularly dissonant. In the presence of these demands, the inability to move to one's own rhythm and to interact at will was translated into objection: "I don't like the ringing sound, it's kind of pushy, you have an obligation to pick up the phone, and feel guilty if you don't" (Mary, Connected Homes 2004–2010).

The persistent disruption characteristic of polychronic dissonance was of course not limited to the home; it was, if anything, even more evident in workplaces, which folded back to affect domestic life. We thus observed a "can't work at work" phenomenon that compelled people to take work home, to escape the office to get work done, or, in some cases, to escape both work and home (and thus media technology) to find periods of uninterruptible time: "I don't get much work done in the office ... I need quiet, and chunks of time ... which I found I just wasn't managing to do in the normal course of either office hours or at home" (June, Connected Homes 2004–2010).

Alternatively, many of our participants spoke of enjoying the dissonant rhythms of digital domesticity and appreciating distraction; interruption; and skipping between tasks, between applications and media, and between work and leisure. Some appreciated the dissonant rhythms that inhered when flicking channels, following

YouTube recommendations, doing a bit of work, social networking, cataloguing family photos, and trawling online news.

> I tend to do a lot of personal internet tasks on work time. I look up online newspapers daily, that is the biggest time waster... I am pretty good. If I have to switch on to work I will, but I am very distracted though if I want to be... there is this online community which I am a part of and I enjoy. Because I work from home there is no one to chat to. (Neil, Connected Homes 2004–2010)

An Asynchronous Consonance

Strategies to manage encounters with digital media in the face of the aforementioned polyphonic drone and polychronic dissonance, and to achieve more consonant (or harmonious) rhythms, often revolved around using the asynchronous affordances of some elements of the household media ecology. Lee and Leibenau describe the use of asynchronous technologies, such as email and texting, as a method to establish patterns of busyness and quiet by accumulating messages "to take advantage of the opportunity to condense or disperse our working effort... and become 'busy' with them at a chosen time" (2002, 270). And, indeed, many of our participants were mindful of the different temporal demands made by different elements of the ecology:

> With email, you can take your own pace. You can never expect to get an immediate reply by email. Real-time [messaging], you have to wait for the other one. It is not simultaneous. Conversation is simultaneous. But you have to dedicate yourself to the communication time and space. (Julian, Connected Homes 2004–2010)

In addition to ordering rhythms through the use of asynchronous communication technologies, many participants imposed ordered routines through asynchronous practices of communication. In their study of mobile phone use, Wajcman et al. (2007, 5) describe asynchronous practices as a common technique involving things like "turning off your mobile to avoid being disturbed." Similarly, in our research, asynchronous practices took the form of limiting media use to designated spaces or times. Here, interactions with the ecosystem centered on establishing and maintaining self-determined routines and deciding when and where media were used, rather than allowing the media to decide when and where they was used (Figure 5.2).

Figure 5.2. Technology creep: photograph of computers spreading into shared and multifunctional spaces like the dining room (Connected Homes 2004–2010)

In the space of the home, the conscious placement of media, and the creation of technology-free zones, was a strategy to stave off the steady colonizing of all of domestic life by media technologies. Similarly, carving out technology-free times, over both short and long periods, was a popular technique. This extended beyond turning technologies off to ignoring technologies if they beeped or rang, particularly during significant or valued family events, such as weekends or mealtimes. The imperative demands of media were, in these ways, resisted to carve out moments of peace amid the background rhythm of the polyphonic drone and polychronic dissonance:

> I keep my [wireless] laptop in the home-office ... I use it for internet and emailing, but not much. I would go on maybe every day, for 5 or 10 minutes. And I check my Hotmail and respond to messages. And that's about all ... my mobile stays in my handbag or in the kitchen ... I am very relaxed about it. If someone wants me, I will eventually get around to it. I am very good at replying and checking. (Angela, Connected Homes 2004–2010)

So we saw that people escaped the drone and the dissonance by strategically "turning off" and carving out parts of space and time that were free of media. However, this did not simply result in a domestic rhythm that oscillated between "on" and "off"; rather, while some media were turned off in the evenings or weekends (email, for example), others were turned on (Netflix, for example), and this differential use indicated that specific technologies were sympathetic to the rhythm of particular domestic events, times, and circumstances.

In these ways, the media ecology was reconciled with and absorbed into routinized or habitual patterns of domestic life that continued, in some ways at least, to take shape around traditional daily, weekly, and yearly routines of working, sleeping, eating, celebrating, and so on. As Eviatar Zerubavel (1985, 4) notes, traditional routines continue to "impose a rhythmic 'beat' on a vast array of major activities (including work, consumption, and socialising)." Similarly, Michael Flaherty and Lucas Seipp-Williams (2005, 43–46) conclude that, despite 24/7 access, the volume and intensity of communication "corresponds to the traditional rhythms of the working day" and that "rhythms in the volume of e-mail are related to other forms of periodicity," including "the biological need for regular sleep coupled with the sociological need for interpersonal synchronization."

These endogenous and traditional temporalities promoted a consonant structure and orderliness to the everyday. For example, weekends and evenings—the times more traditionally associated with family and leisure—were often described by our participants as slower, as less connected, as oriented toward family and shared media rituals (Couldry 2005), and as deliberately insulated from disruption. The rhythms of interaction with the media ecology did not simply displace other rhythms of interaction; rather, they mapped onto or layered them, and, in contrast to technologically determinist rhetoric (prevalent in public discourse), our research suggested that the rhythm of domestic life continued to respond to long-established social norms, biological demands, and other traditional socio-temporal structures, and not in relation to media alone.

But a consonant rhythm could be difficult to maintain, and the entanglements of competing requirements—personal, spatial, biological, social, familial, technological—were not always harmoniously reconciled. For example, a mother and her son discussed arrangements concerning the table in the open-plan living area, which was adjacent to the family television, and was often used by the mother to do work on her laptop computer:

TESS: I tend to work in here, I like working in here. The study is a bit cluttered, and I haven't got time to organise the space ... I like this space, there is nothing here, and I just like [seeing] the backyard.

JAMES: We have had some great arguments about this one. Me sitting here wanting to watch TV, and Mum working at the table with the laptop working saying, "No, don't turn the TV on, I am working, go to the study." Now there is a big TV here, and that thing [laptop] moves, the TV doesn't, it's just illogical ... it drives me insane ... the laptop moves.

TESS: There are some things that drive me insane too ... you haven't actually asked me if it's appropriate to turn the TV on.

JAMES: Of course it's appropriate, the TV is here.

TESS: Maybe I am in the middle of something really important.

JAMES: How can it be that important? It moves.

TESS: Well ask! (Connected Homes 2004–2010)

Orchestrating Domesticity

Our findings echoed Jon May and Nigel Thrift (2001, 12) in suggesting that "rather than a simple picture of speed and acceleration then, the picture that emerges is one of growing awareness of living within a multiplicity of times, a number of which might be moving at different speeds and even in different directions"—and, in the family context, these different speeds and directions need to be orchestrated.

Temporal technologies, such as time charts, schedule books, calendars, and personal organizers of various kinds, have been developed over centuries to facilitate, map, and orchestrate individual temporal trajectories across social, work, and domestic life.

Colin Symes's (1999) research into traditional paper calendars notes that they became critical technologies during the latter part of the twentieth century when the temporal environment shifted from highly routinized, structured, and collectively coordinated "industrial time" to a more contingent, desynchronized, and individualized form of what he calls "professional time." As he argues, "there was no need to keep a diary on an assembly line where the only appointments that workers kept were with a machine" (Symes 1999, 372), whereas, in the present, the distributed and less formally controlled temporal patterns of everyday life mean that chronological latitude has increased, yet simultaneously personal temporal management has become more urgent. In this context, technologies for keeping track of one's schedule facilitate "the meticulous employment of productive chronological habits" (Symes 1999, 360), to establish and maintain a consistency or immutability of chronological ordering across the social (Latour 1990), and to render time more tractable. However, Elizabeth Shove (2008, 5) argues, "[T]he paradox is that systems and devices that promise to increase autonomy and allow individuals greater discretion over the timing and scheduling of activity will, if successful, generate multiple idiosyncratic schedules which in turn increase the problem of coordination."

In the domestic context, the challenge of orchestrating one person's ensemble of routines, activities, events, and people is exacerbated by the fact that, in the "always on, always connected" ecology, others' schedules are also less stable and ordered, exposing the inadequacy of techniques that address personal organization and not group organization. The problem of orchestration extends beyond organizing oneself, beyond individual and personal planning (and personal productivity—Mackenzie 2008), to the necessity of negotiating and organizing one's schedules in relation to others' schedules.

In the contemporary home, we note that this orchestration of articulated movements, events, and requirements employed technologies that had their origins as group-management and project-management applications in the workforce. Predating this, the desire for orchestration was there and was met by older media. One family in our research used a calendar on the fridge that functioned not only as a technique for communicating with one another but as a mechanism for organizing, visually displaying and synchronizing the variable schedules of the family unit:

> We have a family calendar on the fridge... I have a diary and Tom has his Outlook diary. We generally put what all of us are doing on the main one—on the fridge... we have "weekly meetings"—we have to because of Tom's work—to go over what we are doing... so we know what's happening in the next week... we have to get together, because we have separate calendars and a combined one . . . we get together and work out what is happening . . . Tom has an electronic calendar on his computer, which he uses for work, I have a written one. I wouldn't want a technology one. I am much better with written, probably because I have always done it, and I just like the feeling of it. I like looking through it, flipping through pages, counting dates. That's what I like. (Georgie, Connected Homes 2004–2010)

This combination of separate planners coming together on a calendar on "public" display reconciled personal preferences for different media, displaying, at a glance, the daily and weekly schedules of the reticular domestic rhythm. While it assisted in the orchestration of familial schedules through a form of chronometry (time measurement) and temporal cartography (visual mapping) in a highly visible location in the home, it was nonetheless limited in terms of access and use: it was, after all, located, static, and handwritten.

Other families orchestrated their schedules through the use of shared calendar systems. In John and Mary's home the Apple iCal system was one such system we observed to be regularly maintained by all family members, with activities color-coded according to category of activity: "family commitments," "household coordination," "John's work," "Mary's work," "birthdays," and so on (Connected Homes 2004–2010). Similarly, Katie and Bob use synchronized electronic calendars to orchestrate their time:

> We have a computer diary system to check what everyone is doing. It all goes in the diary. Bob organized it all... the orange ones are input on the computer in the back room, and that is on the family dot-Mac account, which goes to the Apple server somewhere before coming back here. Katie's work she inputs on her computer upstairs which is connected to her dot-Mac account... and mine similarly for work appointments... a lot of it is specifically about managing time. (Connected Homes 2004–2010)

The multiple convergent *and* divergent schedules of this particular familial composition were synchronized via a server, which populated data entered from separate devices in real time to visually map and provide a comprehensive overview of the collective. Yet this orchestration of collective schedules was not conducted through centralized control; rather, it was improvised through forms of micro-coordination (Ling and Yttri 1999); it was a distributed and unfolding collaboration composed of incremental adjustments between people, techniques, and technologies. This orchestration *and* improvisation suggested that governing the temporal exigencies of family members and their multiple schedules in the present was a much more flexible dynamic, which challenged the efficacy of former individual or diachronic methods of scheduling. However, this flexibility also implied precarity and a demand for further and intensified management. As Shove (2008, 5) writes, "[O]rganising co-presence becomes more demanding as traditionally shared schedules (e.g. of meal times, of the working day, etc.) gives way to a twenty-four-hour flux of possibilities."

These kinds of temporal management techniques exceeded the management of individual sequences of linear events (Fordism), to include orchestrating the aforementioned polychronicity of multiple time durations, variable schedules, and synchronic activities of household members (just-in-time). This relational coordinating was made legible within the mediated home through the efforts to orchestrate the network of patterns of activity, interaction, connectivity, and communication between and beyond household members. Orchestrating technologies thereby appear as both a solution to and a symptom of a shift from a more established, predictable, and structured industrial time to a more flexible, contingent, fragile, and deroutinized time.

Far from depicting the home as a refuge from the outside world, the contemporary domestic landscape being painted by the likes of Apple, Sony, and Microsoft depicts a teenager in her bedroom plugged into a playlist of streamed music from a server in Los Angeles, while using her mobile device to message her friends in the next street and review the day, while keeping half an eye on her Snapchat feed. In the next room her brother has turned off YouTube and is playing an interactive first-person-shooter. He is "Tobar" in this world, a longstanding member of a "clan" that draws members from Japan, the United States, and Hong Kong. Right now, he is supposed to be accessing his study nodule from his Learning Provider—but he'll come to that later as the organizer app on his smartphone sends ever more insistent reminders. His organizer is coordinated with his stepmother's through the household's shared online calendar—something he doesn't particularly like but accepts as necessary if his alternative to Uber is to ferry him around. On the other hand, he hasn't found it necessary to leave the house for quite some time now.

Other members of the household might be fielding work-related emails, downloading spreadsheets and data files to be updated, writing reports, and making and taking national and international calls and Skype sessions, all the while keeping one eye out for the toddler and the other for the football score. The domestic ecology in this picture of contemporary life is thus a place of leisure, a command and control center, a place for production, and a place for consumption. It is a place that is noisy with people and noisy with incoming and outgoing media. In this chapter, we have examined the variety of ways that all of this is negotiated by parents and their children in the everyday activities of working, playing, educating, socializing, and entertaining in home ecologies now characterized by the polyphonic drone, polychronic dissonance, asynchronous consonance, and orchestrated domesticity. In the following chapter, we shift from analyzing how media ecologies are enacted through negotiated practices to consider how digital domesticity can also be shaped in significant ways by varied modes of *not using* household technologies. As with negotiated use, non-use cultivates household media ecologies in myriad ways that render visible the relational, material, and interdependent nature of contemporary digital living.

[6]

NON-USES

INTRODUCTION

As we have explored in previous chapters, domestic media ecologies are clearly shaped by the range of technologies that inhabit the household environment, and by the householders who put these technologies to use as part of their everyday activities and routines. But such ecologies are also shaped by varied modes of *not* using technologies within the home. This chapter departs from the previous chapters' foci on the materiality of media in the twenty-first-century home, and the ways media have steadily accumulated over time, to inform expectations of use and norms of connectivity and maintenance, by drawing attention to forms of non-use. Non-uses emerged in various guises, mediated by contexts of relationality, materiality, and interdependence we encountered through research within the networked home over this century. Here, we repeatedly observed that use and non-use are not simple binaries that arise in the presence or absence of technologies and resources required to use them, nor are they simply a consequence of individual skillsets or lack thereof, and nor do they occur as a consequence of isolated and distinct decisions. Rather, we consistently found that forms of non-use emerged with, through, and around (1) the digital infrastructures of devices, services, and software configuring the media household; (2) the household population, relationships, and practices doing the configuring; and (3) the ways non-use in turn folded back to reconfigure such arrangements and relations of digital domesticity.

It is important to acknowledge that household media non-use—alongside other forms of technology non-use—is not simply a quiescent negation of use, where use is a practice and is thereby worthy of investigation, and non-use is a situation where nothing is practiced and is thereby not worthy or capable of investigation. Just as neutrality is a political position and silence can be eloquent, so, as we discuss in this chapter, domestic non-use is not simply an absence of use but is itself a form of use-practice. This chapter analyzes the technological and cultural contexts of domestic life in which forms of non-use emerge. It will be seen that these forms of non-use cannot be captured by the nodes of a binary; they either occur at a tangent to the use/non-use binary node axis, or they occur within the spaces between use and non-use. The multidimensional forms of non-use discussions are supported by a growing and broader body of literature that challenges a model where *use* unproblematically equates to an embrace of technology, in binary opposition to *non-use*, which equates simply to eschewing technology.

The issue of not using digital technologies has traditionally been treated as a question of inequality and exclusion, measured in terms of a lack of access to a range of resources, such as money, know-how, hardware, infrastructure, education, or experts, and has in this way been framed as a problem of scarcity or deficit to be overcome. In this "deficit model" the normative default is use, and non-use is the Other, a malady to be diagnosed and if possible redressed. Use is framed as desirable and non-users are framed as "underprivileged," or perhaps as "laggards" who, in an ideal world, are "future users" (Rogers 2003), implying that they are users-to-be who are constrained by a lack of resources and have not been able to keep up with accelerating modernity. In comparisons of the developed world and the developing world, the most commonly cited use constraint is the sheer absence of technology and its infrastructure—hence the "one laptop per child movement," such as Facebook's unsuccessful "Free Basics" initiative in India (Bhatia 2016; Solon 2017) or current efforts from multinational technology companies through Alphabet Inc. subsidiary Loon, and projects like internet.org, a partnership among Facebook, Samsung, Ericsson, MediaTek, Opera Software, Nokia, and Qualcomm, to literally fly in wireless networks with high-altitude balloons (in the case of Loon) or drones (in the case of internet.org).

In this model, these "latent" users have unfulfilled needs to use particular technologies, and concomitantly have unfilled desires to be users of technologies, and, according to United Nations forums such as the World Summit on the Information Society, the Internet Governance Forum, and the Special Rapporteur on the Promotion and Protection of the Right to Freedom of Opinion and Expression, should have these needs and desires fulfilled as a fundamental human right. Meanwhile, in the developed world, as new hardware technologies and new software applications have tumbled over one another at an ever-accelerating pace, the resources-deficit explanation for non-use has shifted from material to human resources, and from a spatial to temporal model. Non-users of "successful" technologies (i.e., technologies that are widely used) will become users in time, and, in so doing, will reinforce the success of the technology, by which time newer technologies will be competing for market attention and success, and a new gap between new technologies and their users and their non-users opens up, ad infinitum.

RELATIONS TANGENTIAL TO USE/NON-USE

Within these contexts of abundant media, there is a growing body of research literature that moves beyond the deficit model of non-use, and beyond "use or non-use" exclusivity, to attend to forms and variations of non-use. Research from a number of fields, including media studies, internet studies, and human–computer interaction (HCI), is coming to understand non-use not as an absolute rejection, or as a delayed embrace, but as being characterized by more selective, fleeting, partial, and mutable relations. While such relations may be driven by provisional and situated constraints, they are also driven by personal, social, or political agendas and decisions (e.g., Selwyn 2006; Satchell and Dourish 2009; Park et al. 2013). For example, Christine Satchell and Paul Dourish (2009) note that "the user" has historically dominated HCI research, and while this is a largely imaginary and discursive category (does the user exist?), it has had the unintended consequence of erasing the very existence of the significant cohort of non-users, or relegates non-users to the delegitimized status of "potential" users (Satchell and Dourish 2009).

In contrast, emerging research on non-use highlights a wide spectrum of individuals who may mistakenly be lumped together as a homogenous Other who is in deficit. The use that non-users make of technologies may in fact take a number of forms. For example, non-users may actively resist or refuse particular technologies altogether (Wyatt 2003; Portwood-Stacer 2013), and be "non-users" by virtue of their minimal, compelled, and disinterested uptake of technologies (Selwyn 2006; Satchell and Dourish 2009), or non-use may involve forms of departure or disconnection associated with various internet or social media platforms, while keeping open the possibility of return (Light and Cassidy 2014). Jed Brubaker, Mike Ananny, and Kate Crawford (2016) describe departure as "neither a singular moment nor a linear path. Rather, it is a process in which people execute, negotiate and undo the meaning of leaving, even while keeping open the possibility of return" (385).

In a persuasive argument that resists the use/non-use binary, Kate Mannell (2017) explores how non-use can be enacted by people who might typically be considered "users," in part by drawing on the work of Michel de Certeau and his concept of "tactics"—tactics being the opportunistic, fleeting, and ephemeral actions taken by "the weak" to evade the structures that dominate them. In this analysis, she argues "the weak" are social media users, the "structures that dominate them" are social media's network effects and their demands for connectivity, and the "tactics" are the strategies taken to be unavailable to the networks while still being a member of the network. These more complex approaches demand that we also consider the importance of political, social, cultural, and economic contexts of technology access and use (e.g., Warschauer 2003). While it is important to acknowledge that, in some circumstances, non-use is symptomatic of inequality and exclusion, and problems of access, both scarcity and deficit abound in the global south as well as parts of the industrialized north, the literature has identified other factors that create alternative positions in relation to use and non-use.

In these models, relations with technology are not positioned along a single axis defined by two nodes—one positive, the other negative—nor adequately represented by being positioned as either a user (read as competent, empowered, contemporary, and efficient) or as a non-user (read as being in deficit, incompetent, disempowered, archaic, and inefficient). Much of this nascent interest in forms of non-use is tangential to this use/non-use axis and emphasizes modes of individual reasoning, choice, or agency that fly in the face of the deficit model and a technological imperative and diffusionist model of staggered yet inevitable technology uptake.

Non-use, then, does not operate in a simple binary relation with use but may well involve more nuanced and cyclical relations of both engagement and disengagement, and more complex motivations than those related to resources and expertise. David Banks describes how, even in practices of non-use, users are aware of their actions, and he suggests that use and non-use "tend to describe social interactions more than they do technological status" (2015, 9). This chapter contributes to this wider discussion by drawing attention to the ways domestic media and technology activities are mediated by the relational, material, and interdependent contexts in which use and non-use take place. That is, non-use does not simply occur as an expression of individual choice but instead operates with and through the increasingly dense materialities of household media ecology infrastructures, devices, and services. Building on the previous chapters, which highlighted uneven distributions of expertise and negotiations around the use of household media, we see that, in many ways, non-use is shaped by the material conditions and reactions of household media ecologies.

Relations to technology that are tangential to use/non-use do not fall neatly into either node and vary across the patterns and spaces of home life, as well as between individual household members. Yet, within each household, forms of engagement and disengagement emerge at the margins of relations to technologies—that is, at the margins of technology design, affordance, functionality, adoption, and use/non-use. Drawing on our own empirical work, interpreted through relevant theory and literature, we have observed practices that are neither use nor non-use, or are tangential to the use/non-use axis, and that can be categorized in terms of the following interstices:

- *Partial use* of the affordances of a particular technology (for example, using a Kindle without an internet connection) or partial use of a technology through time (for example, restricting social media use to the weekends)
- *Active resistance* to certain technologies or applications through strategic disengagement from the totality of available technological resources (for example, watching some television, but refusing to watch commercial television; using email, but refusing to check work email while at home; or using the internet but not social media)
- *Passive neglect* or benign disinterest in available technologies (for example, where a potential user is aware of an application and its purpose and is not necessarily antagonistic to it but lacks the motivation to be an active user)
- *Vicarious use*, where agency over technology and its use is enacted by proxy through a third party, or engagement with technology is at one remove (for example, eschewing personal use of a BitTorrent service in favor of asking someone else with expertise to download a movie or install an app, or sitting back and watching someone else play an interactive game)
- *Radical use*, or deliberately subverting design intentions through repurposing technology (for example, using a dysfunctional tablet as a paperweight)

PARTIAL USE

The *partial use* of a technology was the most common form of practice that we identified in our household research, which was neither straightforward use nor non-use. Instead, it describes a form that either purposefully avoids key affordances of a particular technology while making use of others, or has a temporal element, where use is purposefully part time or is avoided episodically or for particular periods. Here we use technological affordances in Peter Nagy and Gina Neff's (2015, 2) understanding of the term, to refer to the "material qualities of technologies and media as being constituted at least partly outside the communicative, mediate, and affective processes of the people who use them."

At each stage of technology development, new avenues appear for non-use as well as use. As we saw in the early part of the twenty-first century, responding to work email at home was a use of technology that had been a possibility for only a limited duration of time, and many people welcomed this emergent possibility. Concomitantly, the opportunity to refuse to respond to work email from home was also a new possibility welcomed by many.

More recently, partiality—in both its senses—was particularly evident in householders' relations with their smartphones. For instance, Andrew, a father of three (High-Speed Broadband 2011–2017), had an iPhone, which he used to make voice calls and send text messages, yet his phone had no apps installed other than those it came with by default, which Andrew did not use anyway. At the time of writing, about 2 million iPhone apps are available through the Apple Store, and about 2.2 million apps are available for Android phones, many of which are free. In the face of this impossibly large plethora of apps and their affordances, almost all of which are passively neglected, all users are partial users of smartphones. No one can possibly be a user of all the affordances offered by these apps, but leaving this overly literal observation aside, few users make the attempt to make anything like full use of the smartphone's affordances, and many make the purposeful decision to make minimal use of their ever-expanding capacities. Rivka Ribak and Michele Rosenthal (2015), for example, report on voluntary use of "feature phones" (a phone with only basic, cumbersome internet capability) or "basic phones" (a phone that only allows for calling and SMS) as opposed to app-loaded smartphones. Following Louise Woodstock (2014, 1983), who has argued that "media resistance constitutes a significant set of behavioral responses to living in a media-saturated world," this choice to resist smartphones is understood by Ribak and Rosenthal (2015) as an expression of ambivalence toward the ubiquity of smartphones and the media-saturated environment they exemplify.

The partial use of the smartphone's many and varied affordances is interesting because it is at odds with popular descriptions of the appeal of Android and iPhone use, in which the phone is lauded as "being more powerful than the computers that sent man to the moon," and capable of anything and everything. As the meme suggests, no matter what the context, "there is an app for that," from making grocery lists, taking selfies, listening to music, navigating urban geography, tracking fitness, connecting for sex, making art, prioritizing your day, or maintaining your friendships. Indeed, the use of the phone for data communication through these apps exceeds the use of the phone for voice communication for most phone users (Roose 2014), and most phone plans, in Australia at least, are currently priced around data usage with unlimited text and voice call usage.

This phenomenon of loading up devices with more and more functions, more and more features, and more and more options was first identified around the turn of the current century and has been referred to as "feature creep" or "software bloat." Prior to this time, in, say, the 1980s, CPU clockspeeds were much slower, storage capacity was much more limited, and software developers were required to provide "elegant," "minimalist" software so that it could perform adequately on the hardware of the day. As these hardware constraints began to fall away through the 1990s, software developers tended to compete by adding more features and functions to applications, such as word processors, regardless of demand for these features, and despite the decline in usability associated with a more crowded and complicated user interface. The contemporary version of feature creep is evident in the app store and is evident on overcrowded phone interfaces, where a huge variety of functions and features compete for our attention, compete for space on our screens, and compete to be used, and help create a form of "media ambivalence" (Ribak and Rosenthal 2015)—media ambivalence that finds form in active resistance, vicarious use, passive neglect, or partial use at best.

As discussed in previous chapters, we observed a period of time in which households gradually evolved from having a limited number of fixed devices to a growing number of mobile devices. Such arrangements can be traced, in part, to this regime of mobile media and its attendant feature creep beginning with the release of the iPhone in 2007 and the iPad in 2010. These product developments stimulated a surge in household media consumption, underpinned by broadband infrastructure and wireless networks. These conditions of mobile mediation in the home were initially characterized by enthusiastic reception and experimentation with the app economy, though over time we saw phones break, malfunction, and fall into disuse, a subject we turn to in the final chapter. But we also saw much more partial engagement with the abundance of mobile phone applications, including ignoring those available by default, legacy apps that had once been used but now sat idle on the home screen, and a fatigue with experimenting with the latest app offerings in the face of the scale of options in app stores.

Practices were also partial insomuch as use was deliberately "part time" or "episodic" and devices or applications that were used at certain times were also deliberately avoided for extended periods of time. As discussed in earlier chapters, earlier in this century we saw forms of partial use involving "no-go zones" and spatial or temporal "switching off." In research we conducted prior to the introduction of broadband, along with fewer homes using wireless routers and mobile devices, we saw partial use emerge through a stricter regime of spatial organization and management around household media (Wired Homes 2002–2004; Connected Homes 2004–2010). During this period, we found attention on the spatial limitation of media was common through the conscious placement of media and the creation of technology-free zones. These practices were enacted in order to stave off the steady colonizing of the domestic by media technologies. In addition to limiting use of technologies to particular spaces, we saw instances of turning technologies off at certain times to minimize interruptions:

> I turn the phone off if I want to rest. I pull out the plug on the other phone [landline]
> ... I pull it out at night so that no one rings me during the night ... people
> will ring you back ... our philosophy is, what is that critical? (Jacka, Connected Homes 2004–2010)

> Our view is that we only want TV in here, and not out there where we are interacting and having dinner. (Mary, Connected Homes 2004–2010)

Following the steady accumulation of multiple and mobile technologies throughout the house and the everyday, we saw a range of activities and responses that built on these older practices of partial use. This accorded with the findings of other, related studies. For instance, Sarita Yardi Schoenebeck (2014) describes the cyclic patterns of overuse and hiatus among people who give up Twitter for Lent. A study by Lee Rainie et al. (2013) for Pew Research found that 61 percent of users surveyed had taken voluntary breaks from Facebook—"Facebook vacations"—that lasted several weeks or more. And a study by Lee et al. (2014) that focused on distraction and fatigue associated with smartphone use found that 59 percent of the sample strategically employed methods such as altering phone settings (using airplane modes), physical separation (leaving a phone at home), efforts of will (setting personal non-use goals), employing intervention software (that is, software to restrict phone usage), and downgrading to a simpler phone or provider plan. Stacy Morrison and Ricardo Gomez (2014) also explore "pushback" by people who still use contemporary media technologies but take active steps to constrain their use—"a growing phenomenon among frequent technology users seeking to regain control, establish boundaries, resist information overload, and establish greater personal life balance" (n.p).

Other examples of partial use emerged where users were executing a strategy to forgo one set of technical affordances in order to maximize the performance of others. In this case, partial use was not motivated by a desire to meet a need for a less techno-centric life, or out of any other particular subjective need felt by the user, but was motivated by a desire to meet the needs of technologies, and thus maximize their performance, often at the expense of other technologies. Technical contingencies rather than human contingencies were at work, such as, for example, turning off "data" or certain features on a smartphone to preserve battery life. Rationing the use of limited download quotas was something many households had to manage as services became more data intensive, but broadband plans set hard and soft caps on available download volumes. Similarly, bandwidth was something we often observed at the forefront of households' decision making, particularly as the household device population grew, demanding more of the infrastructure in processing, streaming, and downloading data, especially for those not on fiber in rural and regional areas (Figure 6.1).

Bandwidth often lagged behind household media consumption practices, and, along with telco internet plans in which download limits applied, partial use of technologies became a strategy for managing and optimizing the use of this resource. Alternatively, we

Figure 6.1. Photograph of rooftop wireless broadband internet connection in rural home (High-Speed Broadband 2011–2017)

found some homes developed workarounds through the configuration of their media ecology—for example, strategically maximizing internet speeds by forgoing the use of wireless routing and instead connecting devices with Cat 5 cabling, even if they trailed untidily along floors and walls, upstairs and into bedrooms, as we saw in our first visit to Joel's shared house (High-Speed Broadband 2011–2017). Here we saw a tension between the value of internet mobility and speed, in which infrastructures of wireless routing and cabling were configured based on different priorities:

> One cable [goes] out to my computer. One cable out and up to Sam's room. One cable goes to Melissa's room. But Melissa has indicated she does not really care about being wired. She has a laptop and it is really inconvenient for her to sort of lug a cable around the bedroom. We are like, okay, but it is here, in a bundle on the ground, if you need it. (Joel, High-Speed Broadband 2011–2017)

We questioned why there were cables wiring the household's computers given we had just been shown a perfectly good wireless router. Joel and Sam explained that the affordance of additional speed by being wired was more desirable, though evidently not for Melissa. She preferred to trade off the narrower bandwidth concomitant with being wireless in order to shift her laptop around her room more readily, whereas Joel and Sam are avid gamers and made the tradeoff between bandwidth and mobility in favor of bandwidth. As a consequence, wireless routing and wired routing were both partially deployed, and high bandwidth or mobility was each partially achieved.

As is the case with active resistance to technology (discussed later in the chapter), partial use can also be a consequence of not trusting the technology. Users were resistant to certain technologies and certain affordances for reasons that derived largely from the place technology took in the user's *weltanschauung*, or worldview. One household participant was acutely apprehensive about surveillance of their digital activities and about the capture and storage of their metadata; as a consequence, their internet-based practices were partial. Common browsers and search engines were not to be used, and the internet was only to be accessed through a virtual private network, using privacy protective browsers such as Tor and search engines such as DuckDuckGo. Jaume's family had a tablet that was only in the house because it came bundled with the purchase of a USB stick—a bundling that is the hardware equivalent of the above-mentioned feature bloat. The tablet was not specifically invited to join the household and was distrusted. It had never been connected to the internet and had never been used to operate the myriad applications available for the tablet, with the exception of an alarm clock and occasional use of casual games such as Candy Crush that came preloaded on the tablet. Similarly, Jaume's family also owned a Kindle that had not been connected to an internet service. His reasoning was that, once the device is internet enabled, Amazon would have "power" over the files stored on the device: "A few years ago [Amazon] accidently removed books from customers' Kindles" (Jaume, High-Speed Broadband 2011–2017). The family preferred to avoid large internet service providers, such as Amazon, out of paranoia of their tracking algorithms and of their capacity to delete people's entire libraries (King 2012), preferring instead to set up "air gaps" for their devices by downloading content to a single authorized device and then transferring the files.

Partial use of digital media in the home thus came in many stripes. Technologies were used, but their features were not used to the full; or they were used some of the time but not all of the time; or some technologies were partially used in order to maximize the performance of other technologies; or partial use expressed selective distrust. Each form of partial use provides necessary nuance to the notion of a *user* or a *non-user* and is mediated by the relational, material, and interdependent contexts of the home.

ACTIVE RESISTANCE

Some people have access to all of the resources needed to embrace any given technology but, as a consequence of their own technology assessments, choose not to engage and may thus be regarded as *active resistors* of particular modes of technology use.

In some cases, this arises as a consequence of a techno-dystopian critique of technologies per se or, more particularly, of industrial and postindustrial technologies. In this critique, some forms of contemporary technologies are thought to have in their very essence (not just in their use) a propensity to exploit nature, damage the environment, oppress labor, alienate us from ourselves and our fellows, and privilege efficiency as a means over more worthy ends. This dystopian philosophy of technology extends its critical assessment from a societal level to a personal level; from large-scale examples, such as Engel's satanic mills, Heidegger's dams, Marcuse's factory production lines, Winner's nuclear reactors, and Singer's factory farms, to the politics of smaller-scale, everyday technologies such as automobiles, cellphones, or social media. For the techno-dystopian, this essentialist critique can apply to the technologies of everyday life, and, just as the factory farm is actively resisted through a farmers' market, the automobile is resisted through the bicycle, and exploitation of the developing world is resisted through fair-trade purchases, so media consumption can be selectively and actively resisted through non-use.

As far back as 2003, Sally Wyatt et al. recognized that while there were those who had never used the internet through lack of necessary resources (the excluded), and there were those who had been internet users but were no longer users through a lack of necessary resources (the expelled), there were also those who had a choice but by choice have never taken it up (resisters), and there were those who had taken it up but then chose to give it up (rejecters). *Resisting* or *rejecting* is an active rather than passive practice not adequately captured by "non-use." In recent years, those non-users Wyatt et al. (2003) referred to as excluded and expelled have been the focus of those with an interest in internet use, ironically, perhaps, because of the fact that they are non-users. Government agencies, for example, have migrated many services and public-facing interactions from the High Street to the internet, and non-users are an obstacle to these "digital by default" policies. The excluded and the expelled are categories of non-user that must be understood by those with an interest in ever-expanding internet use, but so too are the active resisters, or what Wyatt et al. (2003) referred to as resisters and rejecters. For while the "deficit model" provides a simple solution (no doubt too simple) to what has been defined as a problem for the excluded and expelled (the solution is to redress their deficits), problems and solutions for active resisters are not so simple. What is one to do with people who have the means and the wherewithal but willfully refuse to engage in the information society project? While active resistance and willful rejection of technology may be reasonably widespread, it is also selective in that active resistance to technology is generally not directed at all industrial and postindustrial technologies (with notable exceptions such as Ted Kaczynski) but is expressed through resistance to particular technologies and to particular technologies in particular contexts.

In our own research, communication channels, such as email, and digital file formats, such as MP3 music audio files, have also been frequently raised as technologies subject to critical assessment and active resistance—the former perhaps reflecting its ubiquity and volume of use, and the latter as a consequence of critical assessment of the subtle nuances of music reproduction.

For example, householders contributing to our studies during the Wired Homes project (2002–2004) and early years of the Connected Homes project (2004–2010) were doing so at a time when email was fast gaining momentum, and they viewed this medium as impoverished in comparison with the richness of paper-based personal correspondence, and thus something many were determined to resist. Whereas today electronic communication is part of the fabric of social interaction, for these active resisters emails were a disposable nuisance barely tolerated for work-related correspondence, and definitely not welcome in the personal domain. In contrast, handwritten personal letters were of great significance and were sometimes affectionately collected and carefully archived over a whole lifetime. In the typical family archive were the letters from childhood pen-pals, from family living overseas, and, perhaps of most value for this generation, war correspondence from family members serving overseas.

Mary was one such householder contributing to our studies (Connected Homes 2004–2010). For Mary, handwritten personal letters were of great significance and had been affectionately collected and carefully archived over her lifetime. In the family archive were the letters from childhood pen-pals; letters from her eldest son Peter, written on his first overseas trip; letters from Mary's mother, sent to Mary while she was on her first overseas trip; and war correspondence from Mary's grandmother's brothers. This was certainly not the case with email correspondence, whether personal or not, and Mary remarked: "My question is: How could modern technology and communication preserve these things—Brenda at twelve, Mum at the peak of her Mum-ness, a great uncle who got blown apart in France?" (Mary, Connected Homes 2004–2010). Implying that newer technologies such as email and the mobile phone could not deliver on such subtleties, Mary continued enthusiastically:

> The letters capture [my mum's] voice . . . her dagginess and snitchiness—it's priceless, it really is . . . Those letters capture things for me about my life and other people's lives who I haven't even met . . . I just can't think of anything in today's communication where their personalities would be captured. (Mary, Connected Homes 2004–2010)

At the time of the interview, the fact that a handwritten letter was handmade and not machine made imbued it with the subtlety of craftsmanship and the intimacy of touch, and today, the very fact that the handwritten letter is such an inefficient means of communication in comparison to email, and is consequently becoming more and more rare, imbues it with personal value not matched by email (see Milne 2012 on the enduring significance of letters).

Another common target of critical assessment and active resistance we observed was in relation to music formats, where the phenomenon of audiophiles rejecting digital storage and reproductive technologies in favor of older analog alternatives is well known. The twelve-inch vinyl album is a technology with properties many regard as unmatched by digital alternatives—the cover art, liner notes, the selection and ordering of tracks—in addition to the unique quality of its grooved analog form of recording. The valve-powered amplifier is another such technology, and of course the techniques and technologies associated with the experience of live music stand as the gold standard against which reproductive technologies are judged.

Deborah and Donald, a couple we met in 2013 in regional Victoria, engage in a similar act of active resistance, preferring a record player and tape deck over CDs and MP3s. They extend this resistance to digital audio equipment outside of the home. For example, while their contemporary car came equipped as standard with integrated digital sound systems and with a radio capable of picking up dozens of FM and AM channels, these technologies were actively resisted:

> [In the car] we listen to ABC, because it is the only channel we can get. [...] Sometimes we listen to the news on the car radio. We don't always have that. We rip out anything new and put in tape decks. Have tapes everywhere. Tapes are different. The sound is different. (Deborah, Connected Homes 2004–2010)

Active resistance to email or to digital music is distinct from non-use as a failure to adopt. It involves purposeful assessment of and rejection of selected technologies not captured by the notion of non-use. To assess the properties of technologies such as email communications and music reproduction requires assessment criteria and a threshold for an embrace or rejection. In these cases, criteria such as authenticity and warmth confront criteria such as efficiency and convenience, and active resistance may be the outcome.

Newer devices and applications such as Facebook, Instagram, Twitter, Snapchat, Tinder, Netflix, and so on also confront demanding assessment criteria imposed by active resistors, and may not be used as a consequence. For example, perceived reliability of technology on the "bleeding edge" may lead to non-use through delaying acquisitions until the technology has stabilized:

> I am very conservative when it comes to technology. I'm not keen on buying the latest technology, it's a rat race. . . . [T]oday we have product life cycles where technology comes out with bugs, with problems. Better you buy the technology after the second revision. (Jaume, High-Speed Broadband 2011–2017)

A wide variety of factors motivate rejection or resistance of such applications, including privacy concerns (Tufekci 2008; Guo et al. 2014; Baumer et al. 2015), a desire to curb perceived technology addictions (Baumer et al. 2015), disenchantment with a technology per se (Satchell and Dourish 2009; Morrison and Gomez 2014), a desire to avoid distraction and/or create more productive time (Birnholtz et al. 2013; Baumer et al. 2015; Schoenebeck 2014; Sleeper et al. 2015), and, in the case of social media, concerns about, or disinterest in, a form of socializing that is deemed inauthentic (Tufekci 2008; Woodstock 2014; Sleeper et al. 2015). There is also a range of research on rejection of, or resistance to, Myspace (Tufekci 2008), Grindr (Brubaker et al. 2016), Twitter (Schoenebeck 2014), smartphones (Lee et al. 2014; Ribak and Rosenthal 2015), and, most commonly, Facebook (Tufekci 2008; Karppi 2011; Lampe et al. 2013; Portwood-Stacer 2013; Rainie et al. 2013; Baumer et al. 2015).

PASSIVE NEGLECT

Not all acts of assessment and rejection are expressions of hostility toward the rejected technology, however, and, in this sense, resistance may be more a consequence of *benign* or *passive neglect* rather than active resistance. Indeed, passive neglect is probably the fate of most innovations.

In many cases, there is no obvious reason as to why the technology is not successful (as measured by uptake) whereas, in a different time or a different place, comparable technologies are successful. The history of social network technologies is an interesting case in point. The enormous success of Facebook (as measured by uptake) occurred in the wake of unsuccessful or less successful technologies that preceded it, although they were offering essentially the same functionality (e.g., SixDegrees.com). An example of this was observed in our early research examining the design, implementation, and uptake (or not) of a virtual private network that, at the time, was called a "community intranet" (Wired Homes 2002–2004). In this pre–social media era when web applications were for most people still a novelty, a real estate developer (Stonehenge) and its IT partners (Acumentum and Key4IT) went to considerable expense to design and build a community intranet portal from scratch to be provided to all residents moving into a new housing development in Melbourne, Australia.

In the developers' working documents, briefings to residents, and communications with the authors, the technologies were mobilized with a bewildering number of heroic purposes in mind: building community and reversing alienation through social interaction; networking individuals within families, among neighbors, among the wider community, and with the world; linking households with traditional community groups, such as self-help groups, schools, and clubs; providing an interface between households and local businesses and service providers; providing capacity to work and study from home; and future-proofing the home. To achieve this, every room of all homes was wired with Cat 5 cabling (state of the art at the time), all homes were provided with a computer with the community intranet portal installed as the startup screen, and a community representative council was established to manage and promote the portal.

Alas, in a display of passive resistance that many IT entrepreneurs, app developers, and dot-com investors will be familiar with, residents neglected to use the intranet. Like many other digital innovations, the community intranet enjoyed widespread support in principle and was widely regarded as being well designed and constructed, and its particular functions and its overall objectives had the support of residents (Arnold et al. 2003), but, like Michel Callon's (1984) famous scallops, the residents "did not adhere." As was the case with SixDegrees.com, we found that passive neglect arose in the context of a number of considerations.

Many potential users saw it as a solution in search of a problem; as passive neglect of some accelerated further non-use through the network effect; as technologically second-best to non-technological alternatives—in this case chatting over the fence; or even as a bit strange in its very newness:

> Sure, I'd use it if lots of other people were using it... I don't think I'd be the one to start using it first, however. I never like standing out in that way... but if others were, yeah, I'd have no problems joining in. (Silvana, Wired Homes 2002–2004)

Like other digital technologies of the time, such as email, the idea of an intranet was closely associated with work, labor, and the workplace for many people and not seen as something appropriate for the private or local community sphere. People's ideas (or mental models) about the intranet were shaped by this association, which led to benign neglect:

> An intranet? Oh, yeah ... I use one of them at work! I hate it ... it means I can't access the internet. But, yeah, I know what one is—I can speak to our Sydney office and send documents in-house. What exactly would I do on this community intranet? It seems a bit strange. (Neil, Wired Homes 2002–2004)

Ironically, the efforts the developer put into promoting the intranet through town meetings, organized community consultation, and so forth brought many of the residents together in ways it was hoped the technology would. As one resident quipped at one of these town meetings, "Yes, an intranet is all very well, but do we still get free beer and barbeques?" (Wired Homes 2002–2004). While the activities of the developer were not disliked, and while the community intranet was not actively resisted, it did not find a home within the community (Arnold et al. 2003).

Examples of this form of passive resistance were also observed in our other studies, despite marked improvements in software usability and the overall performance of the technologies in the home. We met Stephanie and Peter in 2013 (High-Speed Broadband 2011–2017) Stephanie was a self-described laggard. At our first meeting, Peter had recently purchased a MacBook laptop, which he hoped Stephanie would like to use for her studies. Stephanie also expressed a desire to use the laptop, but after three years of researcher visits, she had yet to do so. Her stated reason for not engaging with the technology was that between working and parenting she hadn't had the necessary time to familiarize herself with the new device.

Passive neglect can also follow in circumstances where a use relationship has initially been established, through either an immediate cessation or a gradual wind-down. Reflecting the broader phenomenon of what social critics point to as excessive materialism, planned redundancy, and super-affluence in the overdeveloped world, during many of our visits to people's homes across the years, householders described hardware devices that were close at hand and familiar yet couldn't recall the last time they were used; icons for apps on smartphones were no longer recognized; passwords to cloud storage facilities were forgotten; software sits on a laptop without being used for years; and iPad minis sit under the couch for weeks:

> I used to bring this [iPad] to school with a little laptop with a little keyboard, but realistically I haven't used it since school ended. I don't know what... I don't think I have a use for it anymore, really. (Juliette, High-Speed Broadband 2011–2017)

For most of us in the developed world, the stuff we don't use far exceeds the stuff we do use, and the stuff we do use is only partially used. Some of the stuff we don't use is because it is actively resisted, but most of it is simply neglected. The difference between active resistance to technologies and passive neglect of technologies is contextual and relational. At the time of writing, ecologies of household media have been transformed by digital media, with legacy species, such as radio and television, lingering around the margins as part of an aggregation of digital media devices, software, and content. Particular applications of these technologies are so numerous that most can be passively neglected, but non-use of the media species as a whole (television, social media, digital music, etc.) requires active resistance.

VICARIOUS USE

Vicarious use is a form of "use by proxy" observed in our domestic media research where householders purposely avoided performing particular tasks while still accessing the benefits derived from others performing these tasks on their behalf. For example, through proxy arrangements, householders who declined to use social media applications themselves as a matter of personal preference sometimes kept abreast of other householders' use of social media, and thus vicariously accessed social media content, such as newsfeeds, memes, or trending tweets. In effect, this arrangement of vicarious use positioned the user of social media as an editor providing selected social media highlights for the non-user.

Providing proxy access to social media content is selective and partial, though we also observed some vicarious users taking these arrangements one step further by sharing accounts. Just as householders may share a joint bank account or a movie streaming service, we found that social media accounts were also shared. For example, Stephanie did not wish to have her own Facebook account, but she used her husband's account to log in and view posts in his name, and even created new posts assuming his name (High-Speed Broadband 2011–2017).

Many people, however, fiercely guarded the privacy of their social media communications (teenagers in particular) and were very possessive about their devices (especially mobile phones), even though these same householders shared a car, a joint bank account, a large-screen television, a desktop computer, Wi-Fi, or a movie streaming service. Households were thus very selective about which technologies are shared and which are not. Some technologies were understood as "owned" by an individual who had exclusive rights over its use and who actively controlled any vicarious use (smartphones and social media accounts being prime examples), whereas other technologies were not often regarded as personal possessions subject to exclusive use (landline phones and large-screen televisions, for example).

If vicarious use is a form of sharing in which access to content and agency over technology are achieved by proxy, *peripheral use* can be described as a closely related form of non-use in which individuals are positioned at the periphery of other people's use of technology, as "voyeuristic users," as it were. Of necessity or by choice, householders were often all in the same room doing a wide variety of things. In a commonly observed household scene, someone might be watching a reality show on TV while another person was using an iPad while also watching TV, while a third person was reading a book and occasionally responding to comments made about the TV show. Or, someone might be playing a console game and would be watched by someone else not interested in actually playing, but interested in the unfolding interaction, or perhaps the game's developing narrative. The variables at play in these vicarious media consumptions and shared interactions around media consumption would seem to include the character of the hardware (book, tablet, TV, game), the character of the media content (exciting, banal), the relationship between the householders, and the level of concentration of each householder on their own media. Depending on the interplay of these variables, vicarious and peripheral relations to technology became evident and use and non-use became difficult to differentiate. Unsurprisingly, circumstances such as these could be troublesome. Adele lived with her boyfriend Tom, who was an avid Xbox player. Because they had just one screen between them, Adele often got caught up watching Tom play, which frustrated her and led to arguments between the pair: "I hate watching the Xbox. Bor-ing!" (Adele, High-Speed Broadband 2011–2017).

Other manifestations of vicarious use arose through divisions of labor and expertise (see Chapter 4). These were often based on gender and age, and a division of technology-related housekeeping tasks among the household was common, sometimes along the lines of perceived competence, or on the basis of personal preference and confidence, and sometimes as one part of a much larger labor-sharing arrangement for all domestic work (Kennedy et al. 2015). At the time that they engaged with our research, John and Mary (Connected Homes 2004–2010) displayed markedly differentiated relationships with the household's technologies, and while direct use and vicarious use might at times be a consequence of transient subjectivities, these preferences were stable for long enough to settle into the various roles that family members assumed, including gendered roles. So, in a relative sense, John's direct use manifested technological competence and was co-constituted by Mary's vicarious use and lack of affinity with technology. At the same time on a subjective level, John's rapture seemed to correspond to Mary's indifference. John tried to engage her in making decisions regarding the technologies, but it was just not her "thing" and she was perfectly content to leave it to John.

We often observed a particular family member taking the role of the designated "digital housekeeper," who would download films and television shows for other family members, or someone was recognized as the expert in operating the smart television. Centralizing downloading though a household digital housekeeper makes sense in the context of an uneven distribution of experience in locating and downloading files, and understanding of file formats, and where the household has limited bandwidth and a limited download cap, scarce resources may be managed and conserved. Having a designated digital housekeeper (see Chapter 4) also makes sense in terms of conserving labor among household members. The vicarious mode of proxy usage described here is clearly shaped by the constraints of the household's media ecology, available media resources and arrangements for the distribution of labor being key variables.

An uneven distribution of experience, skill, and interest often lead to cascading levels of vicarious use. For example, we observed a parent requesting the household's teenager to download a certain show to a USB stick, and when the file was available in this familiar form, the parent provided the show to young children who were familiar with neither downloads nor USB sticks. Similarly, a more technically skilled householder set up a smartphone to automatically back up images to Dropbox, so another householder made vicarious use of Dropbox each time they took a photograph without having any clear idea of what was going on or how it worked (High-Speed Broadband 2011–2017). Users of movie files on USB sticks and images on Dropbox thereby included people who didn't know how to use USB sticks or Dropbox.

This form of vicarious use of digital technologies has a long history and is also reported in other studies going back to the turn of the century:

> [S]elf-described non-users exploit workarounds that allow them to "use" the internet by having email sent and received by online family members and by having others in their home do online searches for information they want. (Lenhart et al. 2003)

The distinction between use and vicarious use is one in which use produces surplus value consumed by vicarious use. Where habituated, it is a structuring arrangement of labor within the household, and structured arrangements of labor are associated with structured relations of difference.

RADICAL USE

The final form of relationship to technology within households, which again is neither strictly use nor non-use, is *radical use*. Radical use does not use the technology in the normal sense of the term, because it deliberately subverts design intentions for use, and it is not non-use as such, because the technology is repurposed and used for other ends.

Repurposing technologies that are redundant in respect of their original function is a common form of radical use located in different cultural and historical contexts. CDs are placed on tables as drink coasters and hung from trees as bird deterrents, Cat 5 cables become washing lines, cathode-ray televisions become fish tanks, and transistors become cyber-punk jewelry. Contemporary technologies are also subject to radical use.

Over the past two decades social networking sites have been a very common example of radically repurposed technology. Early social networking sites, such as Friendster, were premised on the understanding that users would create profiles that represented real people, and this premise remains today in the terms of use of sites such as Facebook. Enforcing these terms of use by deleting accounts thought to be fake is often said to have been the tipping point leading to a mass migration from Friendster to Myspace, and today Facebook's enforcement of its "real name" "authentic identity" policy is controversial. Facebook claims that "authentic identity is important to the Facebook experience, and our goal is that every account on Facebook should represent a real person" (Facebook spokesperson, cited in Thompson 2012), yet an estimated 80 million Facebook accounts are fake (Thompson 2012). Some people have real names that seem fake (there are fifty-eight people called Donald Duck in the US telephone directory), others have fake names that seem real (John Smith is a popular alias), but, with more serious consequences, some users require a fake identity to socialize while remaining safe from violence. For example, in many parts of the world, political activists, religious dissenters, and LGBTIQ users of social media use a social media alias to seek safety. One such fake account was held by a participant called Jake who used it:

> So that I'm not, there is not a part of me, or aspects of my life owned by something else in the years to come. It's very eccentric and paranoid, but it's just the whole thing like, I know people who've been fired from their jobs because of pictures they've had up on Facebook. And just things like that. It's silly having yourself completely exposed to the open world . . . There is no aspect of my real name or my real photo on my site, so when I add someone I've met because I will talk to them lots of times they will spend days just thinking about whether they should add me because they don't know who I am and then they'll add me and still not know who I am until I start talking to them, and like, "oh—it's you." (Jake, Wired Homes 2002–2004)

A more general take on radical use is a widely held view in science and technology studies that technologies find their purpose in use, not on the designer's workbench, and the deployment of technology in ways that make use of the technology, but do not faithfully reflect the use for which the technology was designed, is the rule rather than the exception. Technologies are thus subject to interpretive flexibility, in which their meaning and function for users and for society at large are not fixed in the technology itself, but emerge in the social contingencies of use as well as in design, and as a consequence of this social construction, emerge in ways that are surprising and diverse as well as mundane. In this sense, radical use is not so radical.

Indeed, many standard uses we regard as normal practices began life as nonstandard or unintended forms of radical use. At times, the development of these forms of use, from radical to accepted standard use, has become so complete that we do not recognize how radical the use is compared with the original design intent. Steven Brown and Geoffrey Lightfoot (2002) chart the adoption of email within organizations at the start of the twenty-first century at a time when email was becoming ubiquitous in many organizations. They found that email, originally intended as a communication tool, was being adopted as a technology of accountability and recordkeeping. Email was used to create records of conversations that could be used to hold others accountable, or to shift accountability to others. It was also used to create personal archives of information. The inbox became not only a place to receive communication, but also a place to store information and documents for latter retrieval. For the designers of early email systems, these uses might seem radical, and certainly strange and unexpected. Today, they seem like a perfectly natural use of the technology.

There were a few examples of this form of relation in our observations; the most striking was one household's use of a home fax/scanner machine. While the home fax machine was very occasionally used to send documents as intended by design, its day-to-day use was for document storage, with important papers, including instructions for the fax machine itself, together with documents containing passwords for online accounts, kept in the scanner tray (Figure 6.2).

> CHARLES: This is what I give to my accountant, bank account details.
> INTERVIEWER: Were you scanning that?
> STEVE: No, he just keeps it there.
> CHARLES: But I have scanned some stuff to send to the solicitor. I keep a record of how to scan, PIN numbers. This is how I do it. I know you have to keep it secure. But this is how I do it. (Charles and Steve, High-Speed Broadband 2011–2017)

Charles's radically different use of the fax/scanner machine was only partially prompted by the redundancy of the device, with little

Figure 6.2. Image of scanner used for document storage (High-Speed Broadband 2011–2017)

need for day-to-day use of the fax/scanner machine as a transmission device, making it ripe for repurposing.

Positioned as binaries, *use* and *non-use* are pure forms that have no place other than as an abstraction. To use a technology in this pure form is to fully use the functions provided by the designers in the manner envisaged in the scenarios of use that inspired the design and were built into the manufacture and marketing of the technologies. In a pure form of use, not only are these functions used faithfully, they are exhausted—they are used to the full, and the potential of the technology is fully realized. At the opposing node, non-use of a technology in the pure form of the binary is for people to have no relation to that technology. From the perspective of the non-user, the technology does not exist.

Neither of these nodes adequately reflects the kinds of relationships people have with technology, which we have observed in our household media research, for no existent technology is (fully) used in daily life, and no existent technology is totally without implications for daily life, even for non-users. The model clearly needs to be nuanced, and the most obvious way is to intercede with the form of use we have termed partial. This intercession suggests that between these two pure forms lies the interstitial space in which the technology is used to some extent without its use being exhausted, and is not used to some extent, while still performing some functions in relation to the user. With use at one end, non-use at the other, and partial use in between, this model might be envisaged as a one-dimensional line with integers marking degrees of use from none to full.

However, our household research suggests that this model is also inadequate, and that there are forms of use and non-use that move away from this single dimension and establish themselves operating not along the use/non-use line, but on another dimension, or at a tangent to that line.

Active resistance, passive neglect, vicarious use, and radical use of technologies in the home are not relations that directly use a technology, nor are they relations that do not use a technology. These practices are forms of media relation that are neither use, non-use, nor simply partial use.

The relationality, materiality and interdependence we encountered through research within the networked home over this century are important in this multidimensional way of thinking about domestic media use and non-use. Most forms of use in marketing and design are framed as autonomous relations forged between a user and a technology. While this user–technology relation is the locus of use, use ought to be understood contextually and perceived as an outcome of wider social, economic, and technological relations shaped with and through ecologies of media, including in this context household media ecologies.

In this approach, active resistance to technologies is not simply non-use, which implies quiescence, but is a vigorous critique of and response to selected aspects of the imperatives or norms associated with media technologies. Passive neglect is in a sense less hostile, but it is no less fatal to technologies and to their use than active resistance; in fact, passive neglect accounts for far more technology failures than active resistance, a failure rate that can only accelerate as the rate at which new technologies clamor for attention accelerates.

But this is not to say that active resistance or passive neglect of a technology eliminates that technology from the lives of the resistors and neglecters and renders them non-users. Just as those who use technology by proxy are vicarious users, through other people using on their behalf, or through witnessing use of the technology, so active resisters and passive neglecters remain vicarious users. Vicarious use places non-users in relation to users rather than separating them out, and this relation establishes us all as users. Just as in a gaming household gaming is part of the shared experience of gamers and non-gamers alike, so in a televisual society, television is part of our experience whether we watch television or not. Television (social media, texting, streaming, etc.) is part of the wider media ecology that we inhabit and live in relation to, and to live in relation to contemporary society is to use technologies like television, social media, and the web, whether directly or vicariously.

Our research provides evidence that intensities of non-use vary across the rhythms and spaces of family life within technologically rich houses and communities. Here, non-use is not a question of individual agency, or lack thereof, nor does it occur in isolation. Instead, it emerges in relation to the people, practices, and technical population of households. A better understanding of household relationship with, through, and around media technologies is developed by unpacking in what part technologies are used and are not used and why. We have suggested in this chapter that active resistance to technologies, passive neglect of technologies, vicarious use of technologies, and radical use of technologies are practices of use and non-use that further nuance both research on household media ecologies and our broader understanding of our relationships with media technologies. In the following chapter we extend this discussion of non-use by exploring technological deterioration, breakage, and obsolescence. We consider how households respond to economic forces and technological imperatives to dispose of the old and upgrade to the new through much more messy ecologies of media practice involving relations of attachment, storage, and reuse.

[7]
DISPLACEMENTS

In this chapter, we consider how households displace and reposition older media technologies, through disposal, storage, repurposing, and passing on to others, as they become obsolete, antiquated, or in some way dysfunctional. We find that some media technologies are retained for symbolic or sentimental value although no longer participating functionally in the household media ecology; others are repositioned by being passed on to other family members, especially children; while others are materially repurposed within the wider ecology of household media stuff to make use of limited functionality. These practices of displacing and repositioning are clearly oriented around physical devices and hardware and yet are also informed by the software functionality of these devices or the data stored in them. Attachments are formed with material things, which makes straightforward disposal difficult, while the lingering functionality and use values of media challenge the imperative of consumer electronics to force obsolescence and upgrading. Concomitantly, then, we report on the steady accumulation of unused media and data-storage devices within homes, motivated by uncertainties around data stored on hard drives, electronic waste protocols, and the possibility that the media and devices will find some future life with the domestic media ecology.

Digital Domesticity. Jenny Kennedy, Michael Arnold, Martin Gibbs, Bjorn Nansen, and Rowan Wilken, Oxford University Press (2020). © Oxford University Press.
DOI: 10.1093/oso/9780190905781.001.0001

These household practices of sharing and storing digital technologies are, however, contrasted with the ways some other legacy media technologies are ruthlessly disposed of, as exemplified by the large number of cathode-ray tube (CRT) televisions littering suburban curbsides, evicted from the home to take their place in the long tail of "zombie media" (Hertz and Parikka 2012) continuing to impact on other environments.

In previous chapters, we discussed various moments when broad and sweeping changes to the technological landscape heralded the influx of new devices into the home. External imperatives, such as the switch from analog to digital terrestrial television broadcasting and the change from "plain old telephone service" (POTS) to phone over internet protocol (IP), have been largely forced on Australian households by changes in legislation and the replacement of nationwide infrastructure. Other broader changes observed in the past twenty years have also had an impact on household media ecologies. The changes noted in previous chapters, such as the broad trajectory of gradual change from a limited number of fixed devices (television, home office computer, stereo hi-fi, kitchen radio) that occupied various central and shared locations to a growing number of mobile wireless devices (smartphones, tablet, laptop computers) occupying fluid and more private spaces. As previously suggested, these changes, and the increased potential for new ways of engaging with media, have stimulated a surge in household consumption of devices.

An increase in accumulation has in turn required displacement and repositioning, including disposal and recycling, passing on, selling in secondhand and secondary markets, as well as storing and stockpiling of old media and devices. Displacement is the flipside to adoption, appropriation, housekeeping, and even non-use, discussed in previous chapters of this book, and we examine in more detail each of these displacement practices in this chapter. It can be seen that in the dynamics of the changing domestic media ecology, displacement must be understood alongside replacement, divestment alongside investment, repositioning alongside positioning, and dispossession alongside possession.

Recent accounts of divestment and displacement have strongly critiqued the notion that we live in a consumerist, throwaway society. For example, Nicky Gregson's (2007a) investigation of UK households found that only 29 percent of household goods were thrown away or directed to the "waste stream." The household goods they refer to were everyday consumer objects, such as furniture and furnishings, television and other electronic media toys, beds and bedding, small and large appliances, and books and games.[1] Sixty percent of these common household goods were "moved along" by being given away to friends and family, sold in secondhand markets, or donated to charity. These findings lead to two surprising results. The first is a direct challenge to the commonsense notion that contemporary consumer culture is part and parcel of a "throwaway society." Second, the afterlife of things emerges through "the pervasive presence of second-hand and hand-me-down/around economies in the practices of everyday life in UK households" (Gregson 2007a, 682). Nevertheless, as we discuss below, such practices and economies of reuse are challenged by the more rapid drive to obsolescence found in consumer electronics and digital media technologies.

---

1. The category of household goods does not include food waste, such as scraps and packaging; it also does not include personal hygiene items and items designed for the care of the body.

OBSOLESCENCE

In his 2007 book *Made to Break*, Giles Slade develops economic and cultural critiques of planned obsolescence, which have shown that consumption has been stimulated through the practices of deliberate obsolescence pursued by American manufacturing industries and emulated around the globe (e.g., Packard 1960). Deliberate obsolescence, he argues, comes in three forms: technological, psychological, and planned. Technological obsolescence occurs when products become obsolete through innovation in functionality—black-and-white television broadcasting gives way to color broadcasting, or feature phones to smartphones. Psychological obsolescence occurs as a result of product manufacturers constantly updating their product models, styles, and features in order to create the desire for the "new" and the "fashionable," although the new and fashionable may be no more functional than the old. It is the "mechanism of changing product style as a way to manipulate consumers into repetitive buying" (Slade 2007, 5). For example, mobile phones that may be functional for many years are upgraded every eighteen months as new models with slightly new or upgraded features are released. Finally, there is material or planned obsolescence. This form of obsolescence involves limiting the life of a product by limiting its durability so that it breaks or wears out quickly. This might be to produce the product more "cheaply" or, in the case of digital media, often emerges through the outdated functionality or performance of software, processing, or memory.

Throughout our fieldwork, we witnessed many examples of media and devices being displaced from the media ecology or from the household itself as they were assessed as obsolete, the most large-scale examples being television and the telephone. During the time of our study, analog CRT televisions were almost universally displaced by flat-screen high-definition digital televisions, and copper-cable telephones by optic fiber alternatives. In significant part this technological obsolescence was beyond the household's control, for as the national media ecology changed so the domestic media ecology was required to change. The terrestrial television infrastructure in Australia broadcast analog television from 1956 and digital television in parallel with analog television between the years 2001 and 2010, until the analog network was phased out and finally completely shut down in 2013. It may be true that the newer digital televisions made the older analog televisions obsolete through offering better-quality pictures and sound and more channels, occupying less physical space while offering bigger screen sizes, and consuming comparatively less power, but it is also true that householders had no choice other than to move to digital receivers, whether through a peripheral digital television set-top box or a completely new television.

During these years, it was common, almost a signature of the times, to see analog CRT televisions put out for curbside trash collections or dumped in empty lots, parkland, and so forth (Figure 7.1). These analog devices were not supported by the digital broadcasting infrastructure and, without the purchase of extra devices, such as set-top boxes and the like, could not take advantage of the new high-definition digital formats. They were generally bulky and unwieldy compared to the new flat-screen digital TVs that were quickly becoming affordable.

Similarly, divestments of obsolescent landline telephones occurred in many of the houses we visited during the rollout of broadband internet. Connection to broadband internet came with voice over internet protocol (VoIP) telephony, and copper-cable telephones were phased out across the whole national network. The change in the national media ecology required all households

Figure 7.1. Dumped CRT television (High-Speed Broadband 2011–2017)

to change fixed-line telephones, along with their accompanying ecology of technologies from answering machines, to wireless handsets and base stations, to fax machines and some alarm systems. Typically, these were put in the garbage can or disposed of in local e-waste or local municipal council hard rubbish collections and thus removed from the house entirely. Less often they were donated to thrift shops. We saw no cases of them being handed on to others. Similarly, other technologies that relied on the old copper telephony system became dysfunctional, leading to their disposal: "A fax machine was recently removed from Dennis' office because it no longer works on the phone line" (Field notes, High-Speed Broadband 2011–2017).

It is important to note that these forms of forced dysfunction are often disruptive to households. In the above case, Dennis relied on the fax machine to receive documents for his business, which involved translating texts for a government department. While the department's communication protocols allowed for confidential documents to be sent by fax, their policy did not allow for the documents to be scanned and emailed. Thus, an important line of revenue was no longer available to Dennis (High-Speed Broadband 2011–2017). The cases of the television and the telephone highlight the ways in which obsolescence is not just a function of interlocking practices and devices in the media ecology of the home, but also how the national and, indeed, international network of policies, practices, technologies, and protocols comes home to roost. As Bruno Latour (1993) famously argued in *We Have Never Been Modern*, the global is local at all points.

We also heard many accounts of people deciding to replace their media and devices due to "psychological" obsolescence—that is, replacing still perfectly functional devices with an updated, newer model regarded as more attractively styled. The most notable example of style-driven obsolescence occurred in the mobile phone market, where regular, often incremental updates drive consumption, even in the absence of the kind of "breakthrough" innovative functionality evident in the displacement of analog television by digital television.

When considering obsolescence, it is useful to remember that technologies, media, and devices have *physical, ecological,* and *social* lives, and obsolescence can occur in any one of these domains.

Physically, they age and deteriorate, break, and fail to work appropriately—a phenomenon taken up in more detail shortly. They also become outmoded in the context of the domestic ecology as a whole; while they themselves may continue to function in the way they always have, this mode of functioning is no longer compatible with the rest of the home's media occupants. In this sense, we can think about them as forming "reverse salients" (Hughes 1987) in the media ecology and practices of the home. In the language of science and technology studies, a reverse salient refers to weak links in the chains of networked hardware and software that significantly compromise the capacity of the technology as a whole to function as intended, or to be improved. In this sense, while a device may continue to be serviceable when considered in isolation, it no longer works seamlessly, or at all, with the rest of a changing domestic media ecology.

> I've actually just upgraded phones. It's technically a very similar plan, but my previous phone didn't like internet connections. It had nothing to do with the broadband service, it was more that the phone was incapable. (Elton, High-Speed Broadband 2011–2017)

> Well, the Apple TV gets used a fraction as much as it did before just because it's just not convenient. It used to be that there was a . . . you choose to do something, and it was available immediately, now there's lagging and buffering and a lot of stuffing around and it just doesn't really feel like it's seamless. (Malcolm, High-Speed Broadband 2011–2017)

Finally, as the social lives of householders change, older devices may no longer support new media practices, altered desires, new activities and pursuits, and new forms of work and leisure. Mobility within the home has been a key factor here through the history of our research. For example, home Wi-Fi coupled with mobile devices such as laptop and tablet computers saw a decoupling of internet and media access from the fixed, desktop PC in the home office. As a result, many of the home offices in the households we visited were phased out. In some households, this happened remarkably quickly. For example, in one home in the Connected Homes project (2004–2010), this change happened almost completely between one visit and the next. The household largely moved from fixed workstations in bedrooms connected to modems via Ethernet cables, to largely using mobile computing devices and Wi-Fi. The television that was once the nightly center of family activity became a thing that was only watched on special occasions, such as a weekly family movie night, and several desktop computers had been packed away.

DYSFUNCTION AND REPURPOSING

Dysfunction to the device and the media ecology occurs when things break, and when devices break there are decisions to be made. Mobile phones are dropped down the toilet, hard drives crash irrevocably, screens crack, and so forth:

> The other one just died on us. It wouldn't turn on. We've still got that. There's stuff on there and we can't turn it on. It was just having its temperamental days. In the end, it was just one of those days where you just kept trying and trying and trying. Just

wouldn't turn on. [sigh] We're going to go out and get another one. (Dianne, High-Speed Broadband 2011–2017)

A drink spilled on a laptop computer, a mobile phone with increasingly short battery life, or a flat-screen television with dead pixels are all breakdowns that force a decision to do without a particular device and shift it to a rarely visited site within the domestic media ecology (storage); move it out of the house entirely; reposition and repurpose it to make use of what residual functionality it might retain; repair it to full functionality; or replace it with a new item. Pragmatic considerations are taken into account in making these decisions, but often the breakage is an opportunity for "psychological obsolescence" to kick in, and the item is replaced with something more current, even though this may not be the most cost-effective solution, and even though the newer device may not offer "breakthrough" functionality.

Our participants routinely reported technologies that broke. Sometimes breakage was accidental, and often calamitous: "There's some broken computers up there. Malcolm has really had a bad run of spilling wine and things and whatever on computers" (Nysha, High-Speed Broadband 2011–2017). Sometimes the technology just seems to wear out, or bits break: "My laptop's speaker is broken so it doesn't play sound" (Marwa, High-Speed Broadband 2011–2017). There may be slow degradations in performance: "It'll get worse, but every now and then it will just turn off. I had to push the button about a hundred times before the screen will appear and stuff like that" (Beverly, High-Speed Broadband 2011–2017). At other times, there is a slow decay in device functioning over time: "It's giving me blue screens more frequently than it used to. Now the keyboard dies every month or so. I think it's falling apart" (Carl, High-Speed Broadband 2011–2017). Technologies may just stop working altogether one day: "I used this for a decade and it takes fantastic photos, but it just died" (Eliza, High-Speed Broadband 2011–2017), again with potentially calamitous results: "Yeah, it died. My old one just, yeah, it was bad. It died, I think, two days before an essay was due" (Dawn, High-Speed Broadband 2011–2017).

Devices do not always break down completely, however; rather, they retain partial functionality and are displaced and repositioned within the ecology to continue to be used in more restricted manners or for other tasks. For example, Jeremy noted about his laptop computers, "The laptop I use has only a half-hour battery life. Other ones do not work at all. So not really laptop laptops. Laptops are hard to Frankenstein" (High-Speed Broadband 2011–2017). Jeremy, who liked to tinker with computers, noted that none of his many laptop computers worked very well, and the parts could not easily be reused to rebuild a functioning computer the way a desktop PC can. However, Jeremy found a use for the one laptop that still functioned—it served a purpose as a computer, but not really as a "mobile laptop" because the device was immobilized by its need to be plugged into an electrical outlet (High-Speed Broadband 2011–2017).

In these ways, degraded or in some fashion broken devices are able to live on thorough a kind of repositioning in the household media ecology. Michael reported that he has a laptop, "but the screen is broken at the moment. It floats around or is used in the front office with a monitor" (High-Speed Broadband 2011–2017). Adele has a similar problem and solution: "The laptop has to be in the corner [next to a monitor] because I broke the screen on it" (High-Speed Broadband 2011–2017). Both Michael and Adele's laptops had dead batteries and had ceased to function as "laptop laptops," and both had now adopted the functionality of a desktop computer in the home office, which harks back to earlier forms of computing in the home. Another participant, Marwa, who was technologically adept and enjoys tinkering with computers and other devices, adopted an approach to managing a broken laptop screen: She used an iPad as the screen for her laptop, and logged into her laptop using a remote desktop app. This workaround for a broken screen also had other benefits for her: It enabled "smoother file access and backup" (High-Speed Broadband 2011–2017). While the laptop computers in all these examples had stopped working as they were designed to work, through adaption and adoption within the media ecology of the home, they could continue to have a working life.

Another means we encountered for dealing with old, obsolete, or dysfunctional media technologies in the domestic ecology was through more deliberate repurposing. Repurposing involves appropriating and adapting a technology to reposition it within a new niche to find new uses for it. Although we witnessed some examples of repurposing, it was in our experience done only by people with strong technical skills and a desire to tinker, which accords with the assertion by Sunyoung Kim and Eric Paulos (2011, 2396) that an important barrier to repurposing old technologies to new ends is a lack of accessible means to discover and invent new ideas for reusing old electronic devices. However, among all the households we visited there were some notable exceptions.

When we visited Bob and Katie's household in 2008 we noted a number of old computers and laptops. Bob and Katie were both professionals and had two young children. Bob worked from his home office. In the corner of their living room they had set up a small, cozy entertainment area:

KATIE: That is where we now watch telly to get away from the kids... we have this little spot now.

BOB: That is an old G5 chip [from an Apple desktop computer]. When it started to get too slow, I said to Katie, "it is substantially cheaper to move the computer over there and use it as a TV, rather than buy a large screen TV..." It is set up as an entertainment machine. It has TV on it, but we don't watch it [TV] much. (Connected Homes 2004–2010)

While the old Apple desktop computer had both internet connectivity and access to terrestrial television, Bob and Katie mainly used it to watch DVDs.

Ivan, whom we visited more recently, lived in remote bushland four hours' drive from Melbourne. His hobbies were astronomy and meteorology. At the time of our visits, Ivan had a home office adjacent to the open-plan kitchen and living areas in which he set up an old laptop, which ran as a weather station and had done so for a number of years. He backed up data to a hard drive situated under the desk once per day (High Speed Broadband 2011–2017).

While deliberately repurposing old technologies was uncommon, it remained an important conduit for displacing old media and devices as part of a broader global and ecological ecology of media, or a form of afterlife for consumer products that Hertz and Parikka (2012, 429) refer to as "zombie media," which is "media that is not only out of use, but resurrected to new uses, contexts and adaptations."

Devices that are retained and repositioned in the digital media ecology to serve as workarounds for broken components and dysfunctions, or are more deliberately repurposed, ultimately played a different kind of role in identity construction to those outlined elsewhere in this book. They did not necessarily play a role in the demonstration of competence or gender. Rather, they played out other identities around the expression of consumer values, not so much of environmental conservation (though this is relevant), but more a rebellion against contemporary society's exhortations for a form of accelerated consumerism that is particularly rampant in the digital technologies sector, where the exuberant call for the acquisition of the "the new, new thing" (Lewis 1999) is particularly insistent and seemingly never-ending. "Hanging on to" devices until they became completely nonfunctional was a form of passive resistance to change and to the work required to divest, replace, repurpose, and update devices. Backing up data, transferring contact lists, and learning new commands and interface elements are all part of this work. Finally, these old devices were often comfortable and "lived in," much as an old pair of slippers or an old T-shirt is comfortable and comforting. They were well worn, well integrated, and well adapted to these people's lives and practices. They were not necessarily to be "shown off" or to serve as status symbols; they might not be "state of the art," but they were comfortable, known, and familiar.

A certain level of dysfunction coupled with repurposing workarounds also led to desirable forms of afterlife for household devices. For example, a tablet computer that was no longer up to the tasks demanded of it by a parent, because, say, it only connects to the G2 mobile networks (no longer available in Australia), might be passed along to children in cases where parents don't want them to have telephony and data connection. The tablet computer was still perfectly adequate for connecting to the internet over Wi-Fi, for playing games, and for other entertaining or educational pursuits.

DISCARDING DEVICES

Discarding is never a simple act of getting rid of a piece of technology. Indeed, though we can remove a device from the home, its material traces persist in much wider geographic and temporal circuits:

> A discarded piece of media technology is never just discarded but part of a wider pattern of circulation that ties the obsolete to recycling centers, dismantling centres in Asia, markets in Nigeria, and so forth—a whole global political ecology of different sorts where one of the biggest questions is the material toxicity of our electronic media. Media kills nature as they remain as living deads. (Parikka 2012)

This "undead" quality of media remains visible through their material traces and ecological impacts. Within the human trajectories of electronics marketing, planned obsolescence, short product life cycles, and rapid technology innovation, the media technologies in our homes are, from a human perspective, quickly outdated and superseded, making their way to the landfill or to an electronic waste disposal center. The human-oriented domestic life of these objects is thus extremely brief, perhaps extending to a few years, or decades at most. In contrast, the material non-human life of these objects extends to many thousands of years (Cubitt 2005; Parikka 2011b). As Tessa Leach says of the Kinect, a range of motion sensors that were designed to work with Microsoft's Xbox video game hardware:

> The Kinect's period of existence will likely far outstrip our own. It has already been brought about through the siphoning of

ancient hydrocarbons through an oil rig, the removal of ore from the earth and the refinement of minerals into glass. Even if these processes do not strictly speaking constitute the life history of the Kinect [...], the future existence of the Kinect is certain to also be highly eventful. We cannot be certain of the future of this and other electronic artefacts, but at present it seems likely that they will be abandoned to decay over thousands of years. For almost all of this time the main interactions for the Kinect will be with objects other than individual humans. (Leach 2018, 96)

Discarded waste moves through various conduits or streams of disposal (Reno 2015). As Gregson et al. (2007b) have suggested, discarding practices can involve a whole variety of strategies such as "binning, giving away, passing-on and selling the surplus or excess" (201). We have seen examples of the ways artifacts that are still useful, even partially, are passed on to other people within the household through repurposing and repositioning. On occasion, such artifacts are also given away to others outside the household. They might end up in secondhand markets, such as garage sales (Hetherington 2004) or flea markets, or be sold through Facebook groups and via other forms of social media. Finally, things can be put away, or stockpiled, deferring the decision about their fate, with the possibility that they will be either reclaimed or divested in the future. In this section of the chapter we discuss three broad practices for the divestment of technologies and devices: (1) disposal through a landfill or recycling, (2) selling in secondary markets, and (3) giving away and passing on to others.

In Australia, waste management for "hard rubbish" is organized by the local council at semiregular times or at household request. It involves household goods not suitable for the regular trash left on the curb for collection. Items collected, as recorded by Moreland

City Council (2018) in Melbourne, include things like "white goods," old or broken furniture, timber scraps, boxes of odds and ends, chinaware, window glass, mattresses and bedframes, electrical appliances, computers and TVs, tools, and lawn mowers. In some places, the collection is made by the local government, in others by private contractors. In each case, material that is of any value is recycled and the rest is taken to the landfill. In many places, it is a local tradition for people to try to beat the local government or contractors to the collection to take items of perceived value. Media devices displaced from one household thus may find a new home in another household. Hard rubbish collection forms one of the major conduits for the divestment of electronics for our participants:

> We have a hard-drive that is not used any more. We got rid of an old computer in hard rubbish. It had been used as storage, and the new server replaced the function of the hard drive and old computer . . . Old phones are recycled. (Riley, High-Speed Broadband 2011–2017)

Others made their own trips to the local landfill or threw out small devices, but, in either case, deciding what could and could not be thrown away posed an dilemma. Once an item is put out for collection and taken away, it is irrevocably gone. How to throw it away was also an issue. Many participants in our studies were very eager to recycle their digital devices. Matthew and Natasha highlighted the dilemma faced when trying to decide how to dispose of technologies that have been replaced:

> MATTHEW: There should be, I don't know, a way to return the thing, especially if you are going to upgrade, to take in the old one. I noticed they are starting to do that with

PlayStation, you can take in the old one and they tell you how much discount you get if you buy a new one. But they should do that with all other technologies, especially televisions . . . There's only so many televisions you can have . . . it's too much of a pain in the neck to sell it, take it to cash converters and they're gonna rob you, so you just wait until a friend needs one.

NATASHA: But they won't, because everything is being superseded. They won't want your old DVD player.

MATTHEW: Yeah, Beta. (High-Speed Broadband 2011–2017)

A common obstacle to recycling media electronics was uncertainty about how to go about it—what could be recycled and what couldn't, and where to take it to be recycled—and a lack of motivation to find out and then actually do it, leading to people storing the items indefinitely through procrastination rather than an attribution of value to the items. A second obstacle was a concern about personal data that might reside on the devices and an unwillingness to send it out where other people could find it. Consequently, a standard practice among all participants was to pass along media technologies that were no longer needed but still functioned perfectly adequately. This might occur when parents acquired a new mobile telephone and passed their old handset down to a child. Handing on and handing down also occurred outside of the household, when obsolete devices, particularly televisions, were passed on to friends, neighbors, and relatives.

Others saw value retained in their digital devices and sought to recoup that value through resale rather than holding on to it. For example, Jens and Jenny, a couple in their twenties living in inner suburban Melbourne, were dedicated iPhone owners (High-Speed Broadband 2011–2017). They regularly debated with each other the relative merits of upgrading their iPhones when new models were released. They wanted the most up-to-date models but found it hard to justify the cost of updating too quickly. They also avoided phones on plans, preferring to buy their phones outright. As Jens said, "It is a bit hard to justify an upgrade every time Apple brings out a new model. I like to wait and get a new phone every second time. Wait for a model." An important part of their strategy for upgrading their iPhones was to resell the old model through an online secondhand phone dealership. As Jens commented:

> No, usually we're . . . once we're finished with it because we go through them so quickly, they've still got some retail value. Usually we sell them to whoever on eBay or there's Facebook groups, and that kind of stuff. They never usually go to waste. We usually don't have them long enough to be, "How do we safely dispose of this device?" It's usually gone before it's even out of production. (High-Speed Broadband 2011–2017)

By selling the phone when it is still twelve to eighteen months old, they can recoup a significant portion of its cost.

Antonio was unusual among the participants in our study in the level of organization and planning he brought to managing his technologies:

> I've got a whole cupboard full of every technology box that I have like, PlayStations, or phones, or anything. They're all sitting upstairs for retail value. You can sell them for a lot more with the box and accessories, and stuff like that than just the device by themselves. [. . .] I usually keep the headphones, the cable, and the wall charger all in the box, never to be used. I usually use the ones I've already got. [. . .] I usually keep those in the box and say, "Brand new accessories, the only thing that's been used is the phone." (High-Speed Broadband 2011–2017)

Other participants in our studies were not so organized or overestimated the value of their old technologies and had trouble reselling them. Steve, a film buff, had a huge collection of movies on DVD. Despite regarding his collection as valuable, he had been unable to sell parts of it:

> Yeah. This is the pile [of DVDs] that I want to keep. There's another pile around the corner which I'm not too sure about. I tried to sell a lot on eBay, but nobody wants them. It's old technology. It's either Blu-ray or high-def downloads, basically. (High-Speed Broadband 2011–2017)

Resale of dated technologies can be difficult. In many instances, old media technologies were taken to Savers, a chain store for secondhand goods, or donated to community-run "op [opportunity] shops" or thrift stores. Donating items recognizes that an antiquated item might retain some vestigial value, but it is not worth the time and effort to attempt to recoup it.

SAVING, STOCKPILING, AND STORING

Drawing on 2008–2009 data from the US Environmental Protection Agency and the National Safety Council, Kim and Paulos (2011, 2395) suggest that "three-quarters of all computers ever purchased in the US remain stockpiled in storerooms, garages or basements of homes, and up to 75 percent of obsolete cell phones are stockpiled in drawers." These figures suggest that an important part of the e-waste stream, as far as consumer electronics is concerned, remains latent. Kim and Paulos (2011) suggest four reasons why people stockpile and store old electronic devices such as cellphones and computers: (1) lack of information about how to recycle these devices, (2) the effort and time required to transport e-waste for recycling, (3) attachment or emotional investment in these devices, and (4) the desire to reuse these objects in new ways but lacking the means to do so.

Without exception, the strategy of displacing media devices within the home through storage was a feature of media displacement and repositioning for all of our informants. When old, infirmed, redundant, broken, obsolete, beyond repurposing, not worth recycling, or of no monetary value, or where uncertainty existed on any of these issues, items were stored. Many of the homes we visited featured multiple disused computers, phones, televisions, media players, external drives, and so forth, tucked away in cupboards, drawers, sheds, garages, and attics. Other households discriminated in that they did not store large items like old desktop computers, printers, or televisions but still had old mobile telephones, laptops, handheld games and gaming consoles, and other small media devices tucked away out of sight in the back of desk drawers, cupboards, and similar locations. Many of these items were stored away and faded from awareness, sometimes to a point where on our technology tours (see Chapter 2) the householders were more surprised than we were at the extent of their inventory of disused devices.

The following memo from fieldwork in Tasmania captures well what we experienced in many homes. As we toured around people's homes, old, stray, unused technological bits and pieces appeared and reappeared like half-forgotten memories that fade from awareness but reappear when prompted or evoked by the process of the technology tour. Beverley, a middle-aged mother of two teenage girls, and her daughter Lucy were taking us through their home in a small Tasmanian town:

Around every corner in this technology tour there seems to be some form of media device, to a point where our informants forgot what or how many things they owned, and needed to amend what they previously said in order to account for the number of tech objects in the home. Initially, Beverley said they have four televisions (there are 5), two laptops (there are 3), and three desktops (there are 4). Beverley remarked, "Oh that's an extra TV down there, that's not even plugged in, it has a really fuzzy picture and it's analogue so going to be defunct pretty soon," and, of an old computer in the corner downstairs, "Oh, I forgot about that one [when giving household tech inventory] . . . it's got [Windows] 98 on it I think . . ." Lucy: "That's my old one." Researcher: "Do things get handed down?" Lucy: "Yeah, I never went on it though, it was just sitting in my room." (High-Speed Broadband 2011–2017)

Beverley's home was littered with stockpiled screens and devices. Many were not used, or only used very infrequently. The family had a flat-screen TV in the living room and three older analog televisions (one in the music room/den for her husband Mark, one in the spare bedroom, and an old portable one downstairs that was not really in use). The flat screen was connected to TiVo, a DVD player, and a videocassette recorder. The television in the spare room was connected to a digital set-top box, PlayStation, and a DVD player that wasn't working. They didn't use the PlayStation much, just occasionally to play SingStar. There was also another television—a new one still in the box—that would be installed in the planned future kitchen. They imagined they would use this for news, casual viewing, and cooking shows. The household also had a wireless router and three laptops. Two were bought by Beverley for work and traveling—a small netbook that was upgraded from an older one. One had been inherited by the girls and they used it for schoolwork or social networking, which were the only uses they had of the internet. The third laptop was Mark's and was kept in the den—he used it for things like shopping on eBay. There were four desktops in the house—an old one not in use in the cupboard in one of the kid's rooms ("just taking up space") and three in Beverley's home office. Two of these were used for work and were connected to printers and covered in all sorts of paperwork (and leftovers from lunch!). The last was a really old computer sitting in a corner that the girls sometimes played old games on. One of the computers downstairs had the family's iTunes music library on it and was used to listen to music. There was a fax machine in the kitchen, a Tastel VoIP phone downstairs, and a Telstra landline (POTS) in the kitchen. Beverley said, "With the new kitchen, everything is going into a cupboard and it will all be wired in . . . I am not getting rid of the Telstra phone, because I don't trust the Tastel system enough with a business phone, as it has dropped out a few times" (High-Speed Broadband 2011–2017).

Beverly had certainly tried to divest some of this plethora of domestic media technologies, with some success and some lack of success:

> We just had a huge garage sale and then went to the Sorrell markets and then the dump to get rid of stuff, every old computer, and none of it sold, even though we were just about giving them away. I've got boxes down there of cables and leads and ports and every other little bit that I've bought over the years to get rid of. (High-Speed Broadband 2011–2017)

The interesting thing with this family was how they revealed that the proliferation of media devices and services—the saturation of the home with media—happens steadily and easily if householders don't actively order and consciously consider how technologies are integrated into the home. Further considerations involve questions of reliability, and how things never work in reality as they are meant to—issues that relate to service problems, inconsistency of speed, and also difficulties in connecting different technologies in a smooth and seamless manner. At the end of our interview, Beverley noted that she imagined a future where things would run more smoothly and would be connected and integrated (television and internet)—yet that these things would come with a cost.

The decision to retain or discard an object is not always an easy one and is not necessarily easier if the object is broken or no longer functions. Dilemmas about repairing or disposing abound, procrastination is easy, and items are stockpiled and stored:

> It's dead, so it's not doing anything at all except being dead, but I use it . . . I like to use it for things like . . . Yeah, although the hard drive died, so that's obviously not saving files. Well, it's got a . . . it's an iMac, so I've got . . . The dead machine is a 27-inch iMac, and it's . . . I don't know, got it just before we got married, so it's a 2008 model, or 2009 model machine, so it's getting to the end of its life, I'm just . . . I don't know, I'm just . . .is it worth spending the money to fix it or replace it at this point is my dilemma and I've just been "uming" and "ahing" about it. (Malcolm, High-Speed Broadband 2011–2017)

Things were also kept because people felt they retained some value. Housemates Shawn and Christine discussed a disused DVD player:

Figure 7.2. Broken mobile phone kept due to uncertainty about disposal (High-Speed Broadband 2011–2017)

SHAWN: Maybe. I don't know. I still can't throw out my [broken espresso machine], so the DVD player will take a while.
CHRISTINE: It's not worthless. It doesn't work, but there's still some value. Someone might buy a broken espresso machine. (High-Speed Broadband 2011–2017)

Even Jens and Jenny, who had well-developed strategies for divesting their devices through secondhand markets, had a number of old, obsolete, or broken mobile phones stored away (Figure 7.2). One was an "emergency" phone to use if their main phones were unavailable. Another was obsolete; it had become outmoded and was no longer of any value on the secondhand market. Finally, they discussed an old, broken mobile phone and not knowing what to do with it:

> I don't know what to do with it. [...] Also, because of the battery and stuff, they're all kind of built in. I cannot just chuck it in the bin. It does make it a lot harder to essentially [...] dispose of them in a safe manner. (Jens, High-Speed Broadband 2011–2017)

They continued to discuss various programs that encouraged people to recycle their phones by providing recycling bags and establishing recycling points at retail outlets. However, despite some abstract awareness of these programs, their lack of practical knowledge about them was a barrier to acting.

Emotional attachments to media devices were also evident throughout our fieldwork and motivated storage rather than discarding. In one of the earlier homes we visited in 2006, we met John (Connected Homes 2004–2010, see Arnold et al. 2006b). As discussed in Chapter 4, emotional attachments were played out through, and fed, John's identity as a "tech-savvy" collector involving the repair and display of old media and computing technologies, such as an antique radio and Morse code transmitter. Most of us collect this sort of "precious junk" (Csikszentmihalyi and Rochberg-Halton 1981)—*junk* insomuch as it does not work and no longer has use value or exchange value, but *precious* as historically obdurate ontological markers. John had domesticated old electronic technologies as a form of ontological work and identity performance. His ability to tinker with them, customize them, and display them scaffolds his sense of place in the home and the world. These subject–object relations extended back in time to John's boyhood experiences with Amateur Electronics, marked his presence in the world, and materially persisted into the future.

Our findings diverge slightly from those of Kim and Paulos (2011) in regard to their suggestion that although people "may indeed desire to reuse obsolete electronics in new ways" (2396), there is no easily available mechanism for them to do so. As described in this chapter, numerous informants repurposed devices and used workarounds to reuse obsolete and partially malfunctioning technologies. We also found that our participants often stored and stockpiled items, not so much in the hope of future innovative or repurposed use (though some did have that possibility in mind) but rather fearing that the devices might be needed at some point in the future, although they were not needed now. For example, getting rid of the videocassette player can create problems for viewing and preserving home videos recorded

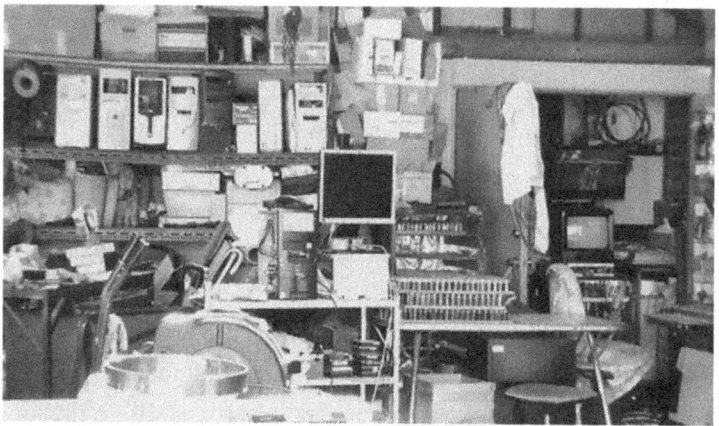

Figure 7.3. Craig's garage, with piles of retained technology, including every mobile phone and every computer ever owned (High-Speed Broadband 2011–2017)

on VHS tapes. Replacing a computer can mean that peripheral devices, such as printers or hard drives, don't work as a consequence of compatibility problems with newer operating systems and software drivers. As Hetherington (2004) suggests, these possibilities are like a "haunting," such that devices can never be fully disposed of, as they are retained as memories and possibilities of use in the future. The feelings about retaining and storing old, unused, and often dysfunctional electronics are also well supported in the literature on material culture, especially the material culture of the home (Csikszentmihalyi and Rochberg-Halton 1981; Rybczynski 1986).

Some of our households reached extremes in terms of stockpiling and storing media and devices. These homes were cluttered with various bits and pieces of electronics and media devices stacked and piled in numerous places around the home. To give one example, Michael's home office had several computers sitting side by side on an old table:

> They're personal. Yeah, that's my old computer, and my old-old computer that next to it. Yeah, they're not worth anything. It's not . . . They just sit there, really. [. . .] No. It's just—it's old junk, really. I don't even know why I've got them. (High-Speed Broadband 2011–2017)

Michael, and others, no doubt retained these old unused devices for a variety of reasons, but, when asked, often could not offer a cogent account. Riley, discussing an old unused laptop, said, "It's in the cupboard. It's collecting dust. I think a friend wanted it" (High-Speed Broadband 2011–2017). Then there are things that are kept, taking up space and collecting dust, often without the exact reason being remembered.

Other homes did not appear cluttered or untidy, and the fact that stockpiling was occurring was not visible on the surface, although hidden away were large amounts of old and disused media and devices. The home itself might be free of the clutter of redundant technologies, while the large shed or the garage contained floor-to-ceiling electronics (Figure 7.3).

Nysha, an artist who works in digital media, has large stockpiles of old, redundant computers. When we visited in 2014, she and her young family had recently moved, which often prompts people to get rid of old, redundant, and unused household objects. However, this was not the case for Nysha: All her old laptop and desktop computers were moved along with the family.

> Yeah, there's loads of hard drives and previous art works and old computers as well up in the attic. Hardware that's the only hardware that particular artwork will run on because [...] you literally can't even buy that hardware anymore. There's a fair bit of that stuff hanging around up in the roof. (High-Speed Broadband 2011–2017)

These multiple computers are Nysha's archive of her digital media art projects, and in a sense an archive of her life. These projects only run on the older versions of software on these computers. While they are not accessed very much, these computers, stored in the attic, form an important part of Nysha's biography, her memory, and her sense of self. With similar motivations, one of this book's authors has on an office shelf a 48k RAM Apple II clone complete with a five-inch external floppy drive—because it was the computer used to complete a master's thesis!

At its extreme, this storing and stockpiling may be regarded as hoarding, a widely recognized psychological disorder. Matthew lived alone in a one-bedroom apartment in the middle suburbs of Melbourne. When we visited him in 2006, we found an apartment filled with old electronics and media devices and associated paraphernalia (Arnold et al. 2006a) (Connected Homes 2004–2010). Entering through the front door, one was confronted with a living room filled with haphazard stacks of electronic goods and associated paraphernalia, including computer monitors, televisions, circuit boards, boxes of cables, radios, typewriters, game boxes, keyboards, and other bits and pieces. A narrow path between towering stacks of digital paraphernalia led to Matthew's desk and chair, where a landline telephone, a stereo, and the only functioning computer in the room were within easy reach. Another path led to the bedroom and bathroom, both also similarly stacked with a huge variety of electronic objects. Most of these objects had no function in a technical sense but existed for Matthew as possibilities for the future; possibilities for re-enchantment and for finding new purposes. Matthew didn't leave his apartment very often—perhaps to see a chiropractor for his back, to submit his biweekly unemployment forms, or to go to the supermarket. However, when he did go out, he often returned with some object, device, or other item, such as an old monitor, a telephone, a TV, or a fax machine he had gathered, or "gleaned," from discards found on the streets: "You never know when something will come in useful, and I hate to see things chucked away." Matthew was very competent at assembling and repairing computers, particularly Apple Macintosh computers. The computer on Matthew's desk was an assemblage of parts he had scrounged and cobbled together from multiple discarded computers.

While Matthew's case may seem extreme—and in many ways it is—it is nevertheless one end of a spectrum of collecting and accumulating practices we saw throughout our various studies. All participants throughout all our studies confronted the often-difficult ethical problem of deciding what to keep and what to dispense with:

> Disposal raises normative questions about how one ought to rid oneself of things, including what should be discarded when and where it ought to go. In this sense, making waste is part of what makes us the ethical selves we want to become. (Reno 2015, 559)

In this chapter, we have examined some of the domestic practices relating to managing and discarding obsolete technologies from within the existing ecology of household media. To further this examination, the reader may well consider how the practices described in this chapter fit within the wider political economy of digital media industries and discourses, in which technologies are continually overrun by newer technologies, and households are forced to engage not only with practical issues of obsolescence, legacy devices, and legacy media, and their configurations in ad hoc assemblages, but also with affective concerns about environmental consequences and the political economies of their electronic waste. We have described observations of households repurposing and working around breakages—perhaps the most positive and benign response to device aggregation. Recycling obsolescent devices is perhaps not as positive as repurposing and working around but is preferable in many respects to discarding to a landfill, and to storing and stockpiling. None of this is easy, but if "the new, new thing" is to take its place and position in the domestic media ecology, the old must be displaced. In the following concluding chapter, we briefly return to the broader questions, themes, and theories of materiality and relationality that we have wrestled with in this book. We have demonstrated through the particular, the contextual, and the messiness of household media ecologies the various stages of technology appropriation, maintenance, negotiation, non-use, and displacement that have unfolded and mutated in the early years of the twenty-first century.

CONCLUSION

Tracing the past twenty years of digital domesticities could readily become a story of progress, a story of yet another domain of life transformed by media technologies in ways that are overwhelmingly beneficial. And indeed, for many, such is the case: The commonly experienced isolation of suburban living has been moderated by the myriad of online social groups and activities people now participate in; the cultural homogeneity of some suburbs has been mediated by cultural diversity available online; opportunities to be entertained at home abound; the financial and opportunity costs of communicating with dispersed family and friends have dramatically decreased; and opportunities for the home to be a place of paid work or an educational center are evident. Yet, as the chapters in this book have highlighted, the past twenty years of digital domesticity have also presented a more complicated story.

The media technologies themselves—the dial-up modems, wireless routers, and mobile smart devices—are in their way things of wonder, but they do not perform unproblematically and transparently according to the functions described on the box. These generations of domestic technologies interact with one another, with the householders and their lives, and with the world beyond the home as a socio-technical ecology that is multilayered and recursively interactive, and with effects too complex to be captured by straightforward notions of progress.

The home itself—the things we ask of a home as a "machine for living"—have not stood still these past twenty years either, nor of course have our work lives, ambitions and desires, families, global and local economies, and communities. These fluxes deny a stable baseline from which any suggestion of progress might be measured. Our story of digital domesticities began with the home management and neighborhood communication system called the "Home Brain," though of course other technologies—the telegraph, radio, telephone, and television—all preceded the Home Brain to shape the environment the Home Brain intended to reshape. Like many of its predecessors the Home Brain was self-consciously positioned by its designers as transformative, another historical example in a long line of techno-determinist and utopian projects. But like many of its antecedents (and descendants), it was an unstable imaginary and its heroic ambitions retreated, expanded, and morphed through numerous design iterations as it struggled to find a place in the increasingly crowded, complex, and competitive domestic media ecology. In the end it failed to survive.

Such localized community intranet infrastructures were, as the 2000s progressed, overtaken by larger infrastructural projects—including national networks of broadband and wireless technologies, alongside the rise in global commercial social media platform architectures. These technical and social infrastructures (Sandvig 2013; van Dijck 2013) displaced more local efforts to design for digital domesticity. Increased bandwidth became part of a wider infrastructural understanding and imaginary of the digital home that enabled the multiplication of devices, especially mobile and touchscreen devices.

CONCLUSION

The significant changes in the domestic media ecology included the replacement of desktop computers with laptop computers, the replacement of analog televisions with digital flat-screen TVs, the replacement of wired devices with wireless devices, and of course the emergence of a range of "smart" devices, beginning with smartphones and later smart home devices, from virtual assistants to home automation systems.

While such diffusion of internet infrastructure and mobile media technologies implies a sense of inevitability in configuring household media ecologies as places of connectivity, interactivity, and productivity, as we have found and detailed in this book, the appropriation, management, and negotiation of relations with, through, and around digital media is a much more contextual and complex process, which is neither uniform nor predictable.

The entangled domestic media ecology, which the Home Brain and subsequent technologies sought to occupy, was imagined as a place of leisure, a command-and-control center, a place for production, and a place for consumption. Such imagined ecologies positioned leisure as a bewildering array of integrated, interactive home entertainment devices for music production, video streaming, interactive game playing, chatting with distant others, and so on. The imagined command-and-control center was to access finances, government services, group-integrated calendars, retail purchasing, and the like. While, as a center of production, the home is imagined as a site and source of data, in many ways returning the digitally connected home to its medieval role as a place for the production of goods and services, mirroring the home's longstanding but expanding role as a location of consumption.

Yet, as we have demonstrated through the close empirical analysis of digital domesticities throughout this book, changes at the level of media materiality do not seamlessly synchronize occupants and their practices with imaginaries of domestic dwelling or functionality.

These socio-technologies are not so much *in* the home but form a dynamic vector that constitutes a certain place *as* a home, and certain ever-changing activities as domestic. At the turn of this century the home was a place where the whole family watched a television in the living room; work was done in a home office with a desk and filing cabinet; only the earliest adopters had a desktop computer; school homework was done on the dining room table as the dishes were washed by hand; telephone calls were made from the wired phone in the hallway or the living room; screens had yet to migrate to bedrooms or dining rooms; and no electronic technologies were mobile, save the paint-spattered Japanese transistor radio. Today's reader might well chuckle at these now quaint household media arrangements and at the "progress" that has been made but would do well to remember how powerful they were in their transformation of the experience of dwelling at home.

As we have seen in this book, the home is an historically fluid and continually evolving institution, one now thoroughly mediated by media and technology. Dwelling with broadband, wirelessness, and distributed and mobile media has, as we have observed, made singular notions or characterizations of home redundant, and instead has participated in the constitution of new modes of dwelling. Focusing on the home's performance as a node in a network of connections, and as a dynamic communicative ecosystem, has enabled us to focus on this fluidity. That is to say, this focus enables us to explore how the home is reshaped or reconfigured socially and materially to accommodate multiple technologies; how boundaries are drawn around the home, and how it is patrolled to control entry and egress; how the technologies are distributed within the home; which sociotechnical "spaces" are private and which are communal;

who uses them and who does not, when, and for what, and under what circumstances; how technologies perform to interpolate the home and the homemakers; how technologies are matched to particular purposes in the home; how the technologies interact, compare, complement, and compete with one another; who the communicators, and who the technophiles and technophobes, are; and how technologies are appropriated, domesticated, and ultimately discarded.

The digital domesticity approach to the home adopted in this book extended beyond analysis of specific technologies (say, television or the internet) and processes of domestication to encompass the interrelated media technologies in the home, envisaged as an "ecology." Our ecological approach has been developed and applied to the home in order to consider the ways in which people, domestic artifacts, household architectures, and new technologies are materially, spatially, and temporally woven together to constitute the particular kind of place called home. This approach has been concerned with the life cycles of digital media and communications technologies populating the home, the materiality of domestic architectures and spaces, the interplay of interpersonal relations and domestic technologies, and the extension of these beyond the domestic through the social and political economies of digital industries and services.

In our approach, digital domesticity constitutes a complex, interactive, socio-technical ecology. By attending to the entire domestic ecology rather than a particular technology, we capture the interrelations of technologies with householders and with one another and provide a more comprehensive and materialist perspective on patterns of digital change within and beyond the home.

A key research aim has been to gain access to households in order to develop a better understanding of such digital domesticities, and the "'intimate histories' of how we live with a variety of media" and other technologies (Morley 2007, 204), through examining these technologies in use, in situ. Methodologically, our empirical work drew on and developed a range of innovative ethnographic and digital approaches—such as household interviews, "technology tours," remote data collection via mobile applications, "domestic probes," and participant-generated data collection—to access and include householders as participants in the research process. These methods were applied over the course of this century through three distinct phases of investigation (see the Project Legend at the front of the book), with some overlap in the times of data collection. Each phase was roughly aligned with significantly different "eras" of digital domesticities, from (1) early dial-up internet connectivity, home offices, and desktop computing, to (2) fixed-line internet, mobile devices, and Wi-Fi, to (3) high-speed broadband, streaming services, mobile devices, multiple screening, and ubiquitous connection.

This book traced these recent histories of household media epistemes, analyzing how various home spaces and their associated practices have been remediated, amended, and reconfigured by the changing domestic media and communications ecology over the course of this century. We discussed how practices of dwelling with media increasingly revolve around relations of accumulation, diffusion, and mobility, and how these dynamics intersect with important spaces within the home, including the kitchen, the bedroom, the living room, and the bathroom. In doing so, we explored how inhabitants rationalized and reconciled the accommodation of new media into existing media ecologies.

Extending theoretical and applied research on domestication through the concept of media ecologies, we in turn examined the extensive "articulation work" needed to bring coherence to and management of the domestic media ecology, and the efforts involved in maintaining the proliferating array of digital technologies within contemporary households, with emphasis on the negotiated rhythms and temporal dynamics at play in the domestic media ecology.

In making their own way into domestic media ecologies, technologies establish their niche and are the beneficiaries of human work in terms of investigating options, making decisions, purchasing, locating, setting up, maintaining, and managing, all of which involves relations of power, authority, gender, labor, and expertise. As we discussed in Chapter 4, housekeeping is a key domestic activity to be reshaped by technologies, including emergent forms of labor required to undertake and perform *digital* housekeeping. The work of situating and maintaining technology in the home and assisting others in its use offers a powerful case study of how identities and roles are negotiated in relation to both housekeeping and technology, how these are gendered, and how particular technologies are themselves gendered in their uses and perceived meanings. Expertise, desires, interests, and duties in relation to digital housekeeping are unevenly acquired, work is unevenly distributed, and our observations of the masculinization of digital housekeeping sets it apart from the very common feminization of other domestic work.

Housekeeping responsibilities have been expanded in the digital home, and so too have parental responsibilities and negotiations, as discussed in Chapter 5. Over the past twenty years, anxieties about children's screen time; the perceived addictive nature of casual, mobile, and online games; and growing concerns about the impact of social media use on children's well-being, including issues of bullying, body image, explicit content, or sleep deprivation all combine to form new grounds for the exercise of parental care and responsibility, and for the exercise of children's independence and maturity. We observed that this new domain of domestic concern was in the main negotiated rather than subject to unilateral decree, and that negotiations differentiated in nuanced and often tentative ways between variables such as age, site, application, device, location within the home, motivation, content, and time of day. These negotiations were ultimately about understandings and norms of digital parenting in the home, and their unsettled configurations as media technologies shifted and children aged and pushed back for rights as digital citizens.

A frequent outcome of these and other household negotiations was the selective non-use of domestic technologies, observed not simply as an absence of technology use, perhaps resulting from a lack of resources such as money or expertise, but as an active form of use-practice driven by personal, social, or political agendas. Our observations in Chapter 6 of these domestic non-use use-practices avoid a black-and-white use/non-use binary to include practices at a tangent to use or non-use. Significant among these were a discerning and contingent *partial use* of the affordances of devices and applications—in particular, smartphones; a *resistance* to selected devices, applications, or affordances based on perceived qualities such as environmental damage, or alienation; a *benign disinterest* in selected devices or applications perceived to be boring, not useful, or more trouble than they are worth; use by *proxy* of selected devices or applications, thereby providing indirect access to technologies while reinforcing household cooperation; and, finally, *radical use* of selected devices or applications such as CDs, social networking sites, and old television sets that repurpose and subvert the intentions of the designers.

A comprehensive consideration of digital domesticity needs to attend to the now routine deployment of smartphones, streaming services, virtual assistants, and so on, but, in so doing, should not elide partial uses, active resistance, passive neglect, vicarious use, and radical use of technologies. Our argument here is that, as with use, non-use cultivates household media ecologies in myriad ways that emphasize the relational, material, and interdependent nature of the contemporary, digital, networked home.

Finally, we explored how and when older things—such as technologies that deteriorated, broke, became obsolete, or were no longer valued—were repositioned within the household media ecology. Perhaps the most decisive form of non-use occurs as technologies were displaced from their position in the domestic ecology, and here, too, we found a variety of nuanced strategies and tactics. In some instances, householders make a decision to displace a technology, but in other cases the technology in question became incompatible with the ecological niche it occupied. Over twenty years the dynamics of change in the domestic ecology has seen digital ecologies displace analog ecologies; cloud storage displace hard-drive storage; wireless devices displace wired devices; flat screens displace cathode-ray tube (CRT) screens; smartphones displace wired phones; fiber-optic cable displace copper wires; on-demand subscription content displace scheduled free-to-air content; and mobile devices displace fixed devices. A lack of interoperability and an inability to continue to interact with the rest of the ecology may mean the technology is materially removed and disconnected from the domestic ecology, but this does not necessarily mean that it "dies." Such technologies are often given a second life in other ecologies, by being relocated to other households for example. Alternatively, they may be relocated within the household, and, though disconnected digitally and functionally, remain connected to the household in a symbolic way, as an historical or sentimental artifact. Other displaced technologies are stored rather than disposed of, reflecting ambivalence about the possibilities of their future usefulness, and many households we visited evidence the steady accumulation of devices and applications over decades, with piles of hardware in garages and spare rooms, and devices choked with software and files not used in years.

It is clear in all of the above that the home is a rapidly changing entity entangled in a maelstrom of rapidly changing media technologies, but it is still not clear that this can be characterized as progress. The connected home is thoroughly integrated in its wider social and economic context, but life within the home is increasingly fragmented and individualized. Gendered as it is, housekeeping is a form of labor that is often elided, and the performance of digital housekeeping is also commonly overlooked by narratives of progress through convenient functionality and intuitive interfaces. The functionality promised and often delivered by the domestic digital ecology provides entertainment; access to consumption, work, and education; and communication far and wide, a form of progress that also erases the centuries-old function of the home as a retreat from the wider world and its demands and distractions. Being connected to the world as thoroughly as the contemporary home is also challenges parents and children to negotiate with one another on new fronts that do not make the task of being a parent and being a child any easier. The evolving domestic technology ecology has also been seen to suffer mass extinctions (of analog technologies, for example), a proliferation of minor and apparent meaningless genetic variations (among streaming apps, for example), the routine deaths of formerly robust technologies (such as hard drives), and the tragic deaths of worthy but uncompetitive species (newspapers, for example). We are reminded that evolutionary competition in an ecology does not reward the best in the sense of the most worthy, and is thus not an inherently progressive dynamic, but rewards the technology best able to compete and reproduce itself regardless of any judgment of worth. This in turn leads to the fate of the dispossessed, the dead, the extinct in the domestic ecology, which often pile up in sedimented drifts, awaiting a future generation to ponder the life of the fossilized remains. In all of this and over twenty years, "progress" in digital domesticity is clearly evident and is not to be denied, but is also ameliorated by complex layers of contingencies and situated interactions.

REFERENCES

Aarsand, Pål André, and Karin Aronsson. 2009. Computer Gaming and Territorial Negotiations in Family Life. *Childhood* 16, no. 4: 497–517.

Access Economics. 2010. *Impacts of Teleworking Under the NBN*. Canberra: Access Economics Pty. Ltd.

Advanced Television. 2018. Australia: Average 6.4 Screens Per Home. *Advanced Television*, March 29. https://advanced-television.com/2017/03/29/australia-average-6-4-screens-per-home/

Alberts, Jess K., Sarah J. Tracy, and Angela Trethewey. 2011. An Integrative Theory of the Division of Domestic Labor: Threshold Level, Social Organizing and Sensemaking. *Journal of Family Communication* 11, no. 1: 21–38.

Allon, Fiona. 2001. An Ontology of Everyday Control: Living and Working in the "Smart House." *Southern Review* 34, no. 3: 8–21.

Altheide, David. 1995. *An Ecology of Communication: Cultural Formats of Control*. New York: Aldine de Gruyter.

Apperley, Tom, Bjorn Nansen, Michael Arnold, and Rowan Wilken. 2011. Broadband in the Burbs: NBN Infrastructure, Spectrum Politics and the Digital Home. *M/C: A Journal of Media and Culture* 14, no. 4. http://journal.media-culture.org.au/index.php/mcjournal/article/view/400

Arnold, Michael. 2004. The Connected Homes Project: Probing the Effects and Affects of Domesticated ICTs. In *Artful Integration: Interweaving Media, Materials and Practices, Vol. 2, Proceedings of the Eighth Biennial Participatory Design Conference*, edited by Adrian Bond, 183–185. Toronto: Computer Professionals for Social Responsibility.

REFERENCES

Arnold, Michael, Martin Gibbs, and Chris Shepherd. 2006. Domestic ICTs, Desire and Fetish. *Fibreculture Journal* 9, no 1. http://nine.fibreculturejournal.org/fcj-059-domestic-icts-desire-and-fetish/.

Arnold, Michael, Martin Gibbs, and Philippa Wright. 2003. Intranets and Local Community: "Yes, an Intranet is all Very Well, But Do We Still Get Free Beer and a Barbeque?"' In *Communities and Technologies*, edited by Marleen Huysman, Etienne Wenger, and Volker Wolf, 185–204. Dordrecht: Kluwer Academic Publishers.

Arnold, Michael, Bjorn Nansen, Jennedy Kennedy, Martin Gibbs, Mitchell Harrop, and Rowan Wilken. 2016. An Ontography of Broadband on a Domestic Scale. *Transformations* 27. http://www.transformationsjournal.org/issues/27/07.shtml

Arnold, Michael, Chris Shepherd, Martin Gibbs, and Karen Mecoles. 2006. Domestic Information and Communication Technologies and Subject–Object Relations: Gender, Identity, and Family Life. *Journal of Family Studies* 12, no. 1: 95–112.

Arrighi, Barbara A., and David J. Maume. 2000. Workplace Subordination and Men's Avoidance of Housework. *Journal of Family Issues* 21: 464–487.

Austin, Erica Weintraub. 1993. Exploring the Effects of Active Parental Mediation of Television Content. *Journal of Broadcasting & Electronic Media* 37: 147–158.

Australian Bureau of Statistics (ABS). 1997. *Unpaid Work and the Australian Economy*, cat. no. 5240.0. Canberra: ABS.

Australian Bureau of Statistics (ABS). 2006. *How Australians Use Their Time*, cat. no. 4153.0. Canberrra: ABS.

Australian Bureau of Statistics (ABS). 2009. *Trends in Household Work, Australia*, cat. no. 4102.0. Canberra: ABS.

Australian Bureau of Statistics (ABS). 2014. *Household Use of Information Technology, Australia, 2012–13*, cat. no. 8146.0. http://www.abs.gov.au/ausstats/abs@.nsf/Lookup/8146.0Chapter12012-13

Australian Bureau of Statistics (ABS). 2016. *Characteristics of Employment, Australia, August 2015*, cat. no. 6333.0. http://www.abs.gov.au/ausstats/abs@.nsf/lookup/6333.0Media%20Release1August%202015

Australian Bureau of Statistics (ABS). 2018. *Household Use of Information Technology, Australia, 2016–17*, cat. no. 8146.0. http://www.abs.gov.au/ausstats/abs@.nsf/mf/8146.0

Australian Centre for Broadband Innovation (ACBI). 2012. *Broadband Connected Homes: Opportunities for Developing Broadband Applications and Services*. Canberra: ACBI.

Australian Communications and Media Authority (ACMA). 2007. *Media and Communications in Australian Families 2007*. Canberra: ACMA.

REFERENCES

Australian Communications and Media Authority (ACMA). 2011a. *Digital Australians: Expectations About Media Content in a Converging Media Environment*. Canberra: ACMA.

Australian Communications and Media Authority (ACMA). 2011b. *Television Sets in Australian Households 2011*. Canberra: ACMA.

Baillie, Lynne, and David Benyon. 2008. Place and Technology in the Home. *Computer Supported Cooperative Work* 17, no. 2–3: 227–256.

Banks, David A. 2015. Lines of Power: Availability to Networks as a Social Phenomenon. *First Monday* 20, no. 11. https://firstmonday.org/ojs/index.php/fm/article/view/6283/5117

Baron, Naomi S. 2008. *Always On: Language in an Online and Mobile World*. New York: Oxford University Press.

Bassett, Caroline, Aristea Fotopoulou, and Katie Howland. 2013. *Expertise: A Scoping Study (Working Paper)*. Leeds: Communities and Culture Network.

Baumer, Eric P. S., Morgan G. Ames, Jenna Burrell, Jed R. Brubaker, and Paul Dourish. 2015. Why Study Technology Non-use? *First Monday* 20, no. 11. https://firstmonday.org/ojs/index.php/fm/article/view/6310

Bausinger, Hermann. 1984. Media, Technology and Daily Life. *Media, Culture and Society* 6, no. 4: 343–351.

Bell, Genevieve, Mark Blythe, and Phoebe Sengers. 2005. Making by Making Strange: Defamiliarization and the Design of Domestic Technologies. *ACM Transactions on Computer-Human Interaction* (TOCHI) 12, no. 2: 149–173.

Berg, Ann-Jorunn. 1997. *Digital Feminism*. Senter for Teknologi of Samfum, Rapport nr. 28. Dragvoll: Norwegian University of Science and Technology.

Bergman, Simone, and Liesbet van Zoonen. 1999. Fishing with False Teeth: Women, Gender and the Internet. In *Technocities: The Culture and Political Economy of the Digital Revolution*, edited by John Downey and Jim McGuigan, 99–107. New York: Sage.

Berker, Thomas, Maren Hartmann, Yves Punie, and Katie J. Ward, editors. 2006a. *Domestication of Media and Technology*. Maidenhead: Open University Press.

Berker, Thomas, Maren Hartmann, Yves Punie, and Katie J. Ward. 2006b. Introduction. In *Domestication of Media and Technology*, edited by Thomas Berker, Maren Hartmann, Yves Punie, and Katie J. Ward, 1–17. Maidenhead: Open University Press.

Bhatia, R. 2016. The Inside Story of Facebook's Biggest Setback. *The Guardian*, May 12. https://www.theguardian.com/technology/2016/may/12/facebook-free-basics-india-zuckerberg

Birnholtz, Jeremy, Lindsay Reynolds, Madeline E. Smith, and Jeff Hancock. 2013. "Everyone Has to Do It:" A Joint Action Approach to Managing Social Inattention. *Computers in Human Behavior* 29, no. 6: 2230–2238.

REFERENCES

Bluedorn, Allen C., Carol Felker Kaufman, and Paul M. Lane. 1992. How Many Things Do You Like to Do at Once? An Introduction to Monochronic and Polychronic Time. *Academy of Management Executive* 6, no. 4: 17–26.

Blunt, Alison. 2005. Cultural Geography: Cultural Geographies of Home. *Progress in Human Geography* 29, no. 4: 505–515.

Blunt, Alison, and Robyn Dowling. 2006. *Home*. New York: Routledge.

Blythe, Mark, and Andrew Monk. 2002. Notes Towards an Ethnography of Domestic Technology. In *Proceedings of the ACM Conference on Designing Interactive Systems* (DIS), 277–281. London: ACM.

Boehner, Kirsten, Janet Vertesi, Phoebe Sengers, and Paul Dourish. 2007. How HCI Interprets the Probes. In *Proceedings of CHI '07: Conference on Human Factors in Computing Systems*, 1077–1086. New York: ACM Press.

Bolin, Göran. 2004. Spaces of Television: The Structuring of Consumers in a Swedish Shopping Mall. In *MediaSpace: Place, Scale and Culture in a Media Age*, edited by Nick Couldry and Anna McCarthy, 126–144. London: Routledge.

boyd, danah. 2010. Social Network Sites as Networked Publics: Affordances, Dynamics, and Implications. In *A Networked Self: Identity, Community, and Culture on Social Network Sites*, edited by Zizi Papacharissi, 39–58. New York: Routledge.

boyd, danah. 2014. *It's Complicated: The Social Lives of Networked Teens*. New Haven: Yale University Press.

Brand, Stewart. 1994. *How Buildings Learn: What Happens After They're Built*. New York: Viking.

Brown, Steven D., and Geoffrey Lightfoot. 2002. Presence, Absence, and Accountability: E-mail and the Mediation of Organizational Memory. In *Virtual Society: Technology, Cyberbole, Reality*, edited by Steve Woolgar, 209–229. Oxford: Oxford University Press.

Brubaker, Jed R., Mike Ananny, and Kate Crawford. 2016. Departing Glances: A Sociotechnical Account of "Leaving" Grindr. *New Media & Society* 18, no. 3: 373–390.

Burgess, Jean, and Joshua Green. 2009. *YouTube: Online Video and Participatory Culture*. Cambridge: Polity.

Burns, Alex, and Stephen McGrail. 2012. Australia's Potential Internet Futures: Incasting Alternatives Using a New Technology Images Framework. *Journal of Futures Studies* 16, no. 4: 33–50.

Bybee, Carl R., Danny Robinson, and Joseph Turow. 1982. Determinants of Parental Guidance of Children's Television Viewing for a Special Subgroup: Mass Media Scholars. *Journal of Broadcasting* 26, no. 3: 697–710.

Cali, Dennis D. 2017. *Mapping Media Ecology: Introduction to the Field*. New York: Peter Lang.

Callon, Michel. 1984. Some Elements of a Sociology of Translation: Domestication of the Scallops and the Fishermen of St. Brieuc Bay. *Sociological Review* 32, no. 1: 196–233.

REFERENCES

Carmody, K. 2010. *Narrative, Contingency and the Innovation Journey: A Case Study of The Wired Homes Project*, PhD thesis, University of Melbourne.

Carroll, John. 2003. The Blacksburg Electronic Village: A Study in Community Computing. *Lecture Notes in Computer Science*, no. 3081: 43–65.

Cellier, J.-M., H. Eyrolle, and C. Marine. 1997. Expertise in Dynamic Environments. *Ergonomics* 40, no. 1: 28–50.

Chambers, Deborah. 2016. *Changing Media, Homes and Households: Cultures, Technologies and Meanings*. London: Routledge.

Clark, Lynn Schofield. 2013. *The Parent App: Understanding Families in the Digital Age*. Oxford: Oxford University Press.

Cockburn, Cynthia, and Ruža Fürst-Dilić. 1994. Introduction: Looking for the Gender/Technology Relation. In *Bringing Technology Home: Gender and Technology in a Changing Europe*, edited by Cynthia Cockburn and Ruža Fürst-Dilić, 1–21. Philadelphia: Open University Press.

Cockburn, Cynthia, and Susan Ormrod. 1993. *Gender and Technology in the Making*. London: Sage.

Connell, Raewyn W. 1987. *Gender and Power*. Cambridge: Polity Press.

Connell, Raewyn W. 2005. *Masculinities*, 2nd ed. Sydney: Alan & Unwin.

Conroy, Stephen. 2009. *New National Broadband Network*, April 7. Canberra: Australian Government. http://www.minister.dbcde.gov.au/media/media_releases/2009/022

Couldry, Nick. 2005. *Media Rituals: A Critical Approach*. London: Routledge.

Crabtree, Andy, Terry Hemmings, Tom Rodden, Keith Cheverst, Karen Clarke, Guy Dewsbury, John Hughes, and Mark Rouncefield. 2003. Designing with Care: Adapting Cultural Probes to Inform Design in Sensitive Settings. In *Proceedings of OzCHI2003: New Directions in Interaction, Information Environments, Media and Technology*, edited by Stephen Viller and Peta Wyeth, 4–13. Brisbane: CHISIG.

Crabtree, Andy, and Tom Rodden. 2004. Domestic Routines and Design for the Home. *Computer Supported Cooperative Work* 13, no. 2: 191–220.

Cristia, Alejandrina, and Amanda Seidl. 2015. Parental Reports on Touch Screen Use in Early Childhood. *PLoS ONE* 10, no. 6: 1–20.

Csikszentmihalyi, Mihaly, and Eugene Rochberg-Halton. 1981. *The Meaning of Things: Domestic Symbols and the Self*. Cambridge: Cambridge University Press.

Cubitt, Sean. 2005. *Ecomedia*. Amsterdam: Rodopi.

Cunningham, Mick. 2007. Influences of Women's Employment on the Gendered Division of Household Labor Over the Life Course: Evidence from a 31-Year Panel Study. *Journal of Family Issues* 28: 422–444.

Daly, Kerry J. 1996. *Families and Time: Keeping Pace in a Hurried Culture*. London: Sage.

Darrah, Charles, J. English-Lueck, and Andrea Saveri. 1997. The Infomated Households Project. *Practicing Anthropology* 19, no. 4: 18–22.

REFERENCES

Davis, Hilary, Martin Gibbs, Michael Arnold, and Bjorn Nansen. 2008. From Exotic to Mundane: Longitudinal Reflections on Parenting and Technology in the Connected Family Home. In *Proceedings of SimTechPaper, Presented at SimTech 2008*, November 20–21, Cambridge.

Department of Broadband, Communications and the Digital Economy (DBCDE). 2010. *Impacts of Teleworking under the NBN*. Canberra: Access Economics.

Department of Broadband, Communications and the Digital Economy (DBCDE). 2011a. *Broadening the Debate: Inquiry into the Role and Potential of the National Broadband Network*. Canberra: House of Representatives Standing Committee on Infrastructure and Communications.

Department of Broadband, Communications and the Digital Economy (DBCDE). 2011b. *#au20 National Digital Economy Strategy: Leveraging the National Broadband Network to Drive Australia's Digital Productivity*. Canberra: Access Economics.

Dourish, Paul, and Genevieve Bell. 2011. *Divining a Digital Future: Mess and Mythology in Ubiquitous Computing*. Cambridge, MA: MIT Press.

Downes, Toni. 2002. Children's and Families' Use of Computers in Australian Homes. *Contemporary Issues in Early Childhood* 3, no. 2: 182–196.

Duffy, Francis. 1990. Measuring Building Performance. *Facilities* 8, no. 5: 17–21.

English-Lueck, J. A. 1998. *Technology and Social Change: The Effects on Family and Community*. Paper presented at the COSSA Congressional Seminar, June 19, pp. 1–7. http://www.svcp.org/pdfs/Technology_and_Social_Change.pdf

Etzioni, Amitai. 1995. *The Spirit of Community*. London: Fontana.

Ewing, Scott, and Julian Thomas. 2012. *CCi Digital Futures 2012: The Internet in Australia*. Melbourne: ARC Centre of Excellence for Creative Industries and Innovation.

Ezell, Stephen J., Robert D. Atkinson, Daniel Castro, and George Ou. 2009. *The Need for Speed: The Importance of Next-Generation Broadband Networks*. Washington, DC: Information Technology and Innovation Foundation.

Fabes, Richard A., Patricia Wilson, and F. Scott Christopher. 1989. A Time to Re-examine the Role of Television in Family Life. *Family Relations* 38, no. 3: 337–341.

Farrington-Darby, Trudi, and John R. Wilson. 2006. The Nature of Expertise: A Review. *Applied Ergonomics* 37, no. 1: 17–32.

Fenner, Grant H., and Robert W. Renn. 2004. Technology-Assisted Supplemental Work: Construct Definition and a Research Framework. *Human Resource Management* 43, no. 2–3: 179–200.

Flaherty, Michael G., and Lucas Seipp-Williams. 2005. Sociotemporal Rhythms in E-mail: A Case Study. *Time & Society* 14, no. 1: 39–49.

Flichy, Patrice. 1995. *Dynamics of Modern Communication: The Shaping and Impact of New Communication Technologies*. London: Sage.

Foucault, Michel. 1980. The Confession of the Flesh. In *Power/Knowledge Selected Interviews and Other Writings*, edited by Colin Gordon, 194–228. New York: Pantheon.

REFERENCES

Franklin, Adrian. 2006. "Be[a]ware of the Dog": A Post-humanist Approach to Housing. *Theory and Society* 23, no. 3: 137–156.

Frissen, Valerie. 1995. Gender Is Calling: Some Reflections on Past, Present and Future Uses of the Telephone. In *The Gender–Technology Relation: Contemporary Theory and Research*, edited by Keith Grint and Rosalind Gill, 79–84. London: Taylor and Francis.

Fuller, Matthew. 2005. *Media Ecologies: Materialist Energies in Art and Technoculture.* Cambridge, MA: MIT Press.

Gaver, Bill. 2001. Designing for Ludic Aspects of Everyday Life. *European Research Consortium for Informatics and Mathematics (ERCIM) News* 47 (October): 20.

Gaver, Bill. 2002. *Home Is Heaven for Beginners: Probes and Proposals for Domestic Technologies.* Position paper for "Technology for Families" workshop at Conference on Human Factors in Computer Systems (CHI 2002), April 20–25, Minneapolis. http://www.cs.umd.edu/hcil/interliving/chi02/gaver.pdf

Gaver, Bill, Tony Dunne, and Elena Pacenti. 1999. Design: Cultural Probes. *Interactions* 6, no. 1: 21–29.

Gaver, Bill, and Heather Martin. 2000. Alternatives: Exploring Information Appliances Through Conceptual Design Proposals. In *Proceedings of Conference on Human Factors in Computing Systems CHI 2000,* The Hague, Netherlands 2, no. 1: 209–216.

Gaver, William W., Andrew Boucher, Sarah Pennington, and Brendan Walker. 2004. Cultural Probes and the Value of Uncertainty. *Interactions* 11, no. 5: 53–56.

Glaser, Robert, and Michelene T. H. Chi. 1988. Overview. In *The Nature of Expertise*, edited by Michelene T. H. Chi, Robert Glaser, and M. J. Farr, xv–xxxvi. Mahwah, NJ: Lawrence Erlbaum.

Graham, Connor, Mark Rouncefield, Martin Gibbs, Frank Vetere, and Keith Cheverst. 2007. How Probes Work. In *Proceedings of the 19th Australasian Conference on Computer–Human Interaction: Entertaining User Interfaces (OZCHI '07),* 29–37. New York: ACM.

Graham, Steve, and Simon Marvin. 2001. *Splintering Urbanism: Networked Infrastructures, Technological Mobilities, and the Urban Condition.* London: Routledge.

Green, Lelia, and Donell J. Holloway. 2014. 0–8: Very Young Children and the Domestication of Touchscreen Technologies in Australia. In *Proceedings of the Australian and New Zealand Communication Association Annual Conference.* Melbourne: Swinburne University of Technology.

Green, Lelia, Donell Holloway, and Robyn Quin. 2004. @ Home: Australian Family Life and the Internet. In *Virtual Nation: The Internet in Australia,* edited by Gerard Goggin, 88–101. Sydney: UNSW Press.

Gregg, Melissa. 2013. *Work's Intimacy.* New York: John Wiley & Sons.

Gregg, Melissa. 2018. *Counterproductive: Time Management in the Knowledge Economy.* Durham, NC: Duke University Press.

REFERENCES

Gregg, Melissa, and Jason Wilson. 2011. *Willunga Connects: A Baseline Study of Pre-NBN Willunga*. Adelaide: DEEST.

Gregson, Nicky. 2007. *Living with Things: Ridding, Accommodation, Dwelling*. Canon Pyon, Herefordshire: Sean Kingston Publishing.

Gregson, Nicky, Alan Metcalfe, and Louise Crewe. 2007a. Identity, Mobility and the Throwaway Society. *Environment and Planning D: Society and Space* 25, no. 4: 682–700.

Gregson, Nicky, Alan Metcalfe, and Louise Crewe. 2007b. Moving Things Along: The Conduits and Practices of Divestment in Consumption. *Transactions of the Institute of British Geographers* 32, no. 2: 187–200.

Groves, Derham. 2004. *TV Houses. Television's Influence on the Australian Home*. Melbourne: Black Jack Press.

Guo, Yanru, Dion Goh Hoe-Lian, Nor Aishah Binte Mohamed Rashid, Pyae Pyae Han, and Shengbo Sun. 2014. An Investigation of Twitter and Facebook Abandonment. In *Proceedings of the New Generations (ITNG), 2014 11th International Conference on Information Technology*, April 7–9, Las Vegas, Nevada.

Haddon, Leslie. 1992. Explaining ICT Consumption: The Case of the Home Computer. In *Consuming Technologies: Media and Information in Domestic Spaces*, edited by Roger Silverstone and Eric Hirsch, 82–96. London: Routledge.

Haddon, Lesley. 2003. Domestication and Mobile Telephony. In *Machines that Become Us: The Social Context of Personal Communication Technology*, edited by James E. Katz, 43–56. New Brunswick, NJ: Transaction Publishers.

Haddon, Lesley. 2004. *Information and Communication Technologies in Everyday Life: A Concise Introduction and Research Guide*. Oxford: Berg.

Haddon, Leslie. 2011. Domestication Analysis, Objects of Study, and the Centrality of Technologies in Everyday Life. *Canadian Journal of Communication* 36: 311–323.

Hahn, Marcus, and Roger Wilkins. 2014. Households and Family Life. In *Families, Incomes and Jobs, Volume 9: A Statistical Report on Waves 1 to 11 of the Household, Income, and Labour Dynamics in Australia Survey*, edited by Roger Wilkens, 2–6. Melbourne: Faculty of Business & Economics, University of Melbourne. http://www.ncsehe.edu.au/wp-content/uploads/2014/06/Families-Incomes-and-Jobs-Vol-9.pdf

Hall, Edward T. 1959. *The Silent Language*. Garden City, NY: Doubleday.

Hampton, Keith N. 2000. Grieving for a Lost Network: Collective Action in a Wired Suburb. *Information Society* 19, no. 5: 417–428.

Hampton, Keith N., and Barry Wellman. 2000. Examining Community in the Digital Neighborhood: Early Results from Canada's Wired Suburb. In *Digital Cities: Experiences, Technologies and Future Perspectives*, edited by Toru Ishida and Katherine Isbister, 194–208. Heidelberg, Germany: Springer-Verlag.

REFERENCES

Hand, Martin, Elizabeth Shove, and Dale Southerton. 2007. Home Extensions in the United Kingdom: Space, Time, and Practice. *Environment and Planning D: Society and Space* 25, no. 4: 668–681.

Haring, Kristen. 2003. The "Freer Men" of Ham Radio: How a Technical Hobby Provided Social and Spatial Distance. *Technology and Culture* 44, no. 4: 734–761.

Harman, Graham. 2018. *Object-Oriented Ontology: A New Theory of Everything*. London: Pelican Books.

Harper, Richard, ed. 2011. *The Connected Home: The Future of Domestic Life*. London: Springer.

Hayles, N. Katherine. 1999. *How We Became Posthuman: Virtual Bodies in Cybernetics, Literature and Informatics*. Chicago: University of Chicago Press.

Hearn, Gregory N., and Marcus Foth. 2007. Communicative Ecologies: Editorial Preface. *Electronic Journal of Communication* 17, no. 1-2: 1–2. http://www.cios.org/www/ejc/v17n12.htm#introduction

Heise, Ursula K. 2002. Unnatural Ecologies: The Metaphor of the Environment in Media Theory. *Configurations* 10, no. 1: 149–168.

Hemmings, Terry, Andy Crabtree, Tom Rodden, Karen Clarke, and Mark Rouncefield. 2002. Domestic Probes and the Design Process. In *Proceedings of the 11th European Conference on Cognitive Ergonomics*, 187–193. Utrecht, Netherlands: European Association of Cognitive Ergonomics.

Hertz, Garnet, and Jussi Parikka. 2012. Zombie Media: Circuit Bending Media Archaeology into an Art Method. *Leonardo* 45, no. 5: 424–430.

Hetherington, Kevin. 2004. Secondhandedness: Consumption, Disposal and Absent Presence. *Environment and Planning D: Society and Space* 22, no. 1: 157–173.

Highmore, Ben. 2010. *Ordinary Lives: Studies in the Everyday*. London: Routledge.

Hine, Christine. 2000. *Virtual Ethnography*. London: Sage.

Hitchings, Russell. 2004. At Home with Someone Nonhuman. *Home Cultures* 1, no. 2: 169–186.

Hommels, Anique. 2005. Studying Obduracy in the City: Toward a Productive Fusion Between Technology Studies and Urban Studies. *Science, Technology, and Human Values* 30, no. 3: 323–351.

Hourcade, Juan Pablo, Sarah L. Mascher, David Wu, and Luiza Pantoja. 2015. Look, My Baby Is Using an iPad! An Analysis of YouTube Videos of Infants and Toddlers Using Tablets. In *Proceedings of CHI 15*, 1915–1924. New York: ACM Press.

Hughes, Thomas P. 1987. The Evolution of Large Technological Systems. In *The Social Construction of Technological Systems*, edited by Wiebe E. Bijker, Thomas P. Hughes, and Trevor Pinch, 51–82. Cambridge, MA: MIT Press.

Inkster, Ian. 1991. *Clever City: Japan, Australia and the Multifunction Polis*. Oxford: Oxford University Press.

Jankowski, Nicholas W. 2002. Creating Community with Media: History, Theories and Scientific Investigations. In *Handbook of New Media: Social Shaping and*

REFERENCES

Consequences of ICTs, edited by Leah A. Lievrouw and Sonia Livingstone, 34–49. London: Sage.

Johnson, Lesley, and Justine Lloyd. 2004. *Sentenced to Everyday Life: Feminism and the Housewife*. Oxford: Berg.

Jones, Steven G. 1997. The Internet and its Social Landscape. In *Virtual Culture: Identity and Communication in Cybersociety*, edited by Steven G. Jones, 7–35. London: Sage.

Karppi, Tero. 2011. Digital Suicide and the Biopolitics of Leaving Facebook. *Transformations: Journal of Media & Culture* 20. http://www.transformationsjournal.org/wp-content/uploads/2016/12/Karppi_Trans20.pdf

Kennedy, Jenny, Michael Arnold, Bjorn Nansen, Rowan Wilken, and Martin Gibbs. 2015. Digital Housekeepers and Digital Expertise in the Networked Home. *Convergence* 21, no. 4: 408–422.

Kim, Sunyoung, and Eric Paulos. 2011. Practices in the Creative Reuse of e-Waste. In *CHI'11, Proceedings of the SIGCHI Conference on Human Factors in Computing Systems*, Vancouver, BC, Canada, May 7–12, 2395–2404.

King, Mark. 2012. Amazon Wipes Customer's Kindle and Deletes Account With No Explanation. *The Guardian*, October 23. www.theguardian.com/money/2012/oct/22/amazon-wipes-customers-kindle-deletes-account.

Knorr-Cetina, Karin. 1997. Sociality with Objects: Social Relations in Postsocial Knowledge Societies. *Theory, Culture and Society* 14, no. 4: 1–30.

Koolstra, Cees M., and Nicole Lucassen. 2004. Viewing Behavior of Children and TV Guidance by Parents: A Comparison of Parent and Child Reports. *Communications* 29, no. 2: 179–198.

Krajina, Zlatan, Shaun Moores, and David Morley. 2014. Non-media-centric Media Studies: A Cross-Generational Conversation. *European Journal of Cultural Studies* 17, no. 6: 682–700.

Lachance-Grzela, Mylène, and Geneviève Bouchard. 2010. Why Do Women Do the Lion's Share of Housework? A Decade of Research. *Sex Roles* 63, no. 11–12: 767–780.

Lally, Elaine. 2002. *At Home with Computers*. Oxford: Berg.

Lampe, Cliff, Jessica Vitak, and Nicole Ellison. 2013. Users and Nonusers: Interactions Between Levels of Adoption and Social Capital. In *Proceedings of the 2013 Conference on Computer-Supported Cooperative Work*, February 23–27, San Antonio, Texas.

Latour, Bruno. 1990. Drawing Things Together. In *Representation in Scientific Practice*, edited by Michael Lynch and Steve Woolgar, 19–68. Cambridge, MA: MIT Press.

Latour, Bruno. 1992. Where are the Missing Masses? The Sociology of a Few Mundane Artifacts. In *Shaping Technology/Building Society*, edited by Wiebe E. Bijker and John Law, 225–258. Cambridge, MA: MIT Press.

REFERENCES

Latour, Bruno. 1993. *We Have Never Been Modern*. New York: Harvester Wheatsheaf.
Latour, Bruno. 2002. Morality and Technology. The Ends of the Means, trans. Couze Venn. *Theory, Culture & Society* 19, no. 5–6: 247–260.
Latour, Bruno. 2005. *Reassembling the Social: An Introduction to Actor-Network Theory*. Oxford: Oxford University Press.
Law, John. 2004. *After Method: Mess in Social Science Research*. London: Routledge.
Law, John, and John Hassard. 1999. *Actor-Network Theory and After*. Oxford: Blackwell.
Leach, Tessa. 2018. *Anthropomorphic Machines: Alien Sensation and Experience in Nonhumans Created to Be Like Us*. Unpublished PhD thesis, University of Melbourne.
Lee, Heijin, and Jonathan Liebenau. 2002. A New Time Discipline: Managing Virtual Work Environments. In *Making Time: Time and Management in Modern Organizations*, edited by Richard Whipp, Barbara Adam, and Ida Sabelis, 126–139. New York: Oxford University Press.
Lee, Hongsub. 1999. Time and Information Technology: Monochronicity, Polychronicity, and Temporal Symmetry. *European Journal of Information Systems* 8: 16–26.
Lee, Uichin, Subin Yang, Minsam Ko, and Joonwon Lee. 2014. Supporting Temporary Non-Use of Smartphones. *CHI14: Refusing, Limiting, Departing: Why We Should Study Technology Non-Use Workshop*, April 26–May 1, Toronto, Canada, 65–68.
Lefebvre, Henri. 2004. *Rhythmanalysis: Space, Time and Everyday Life*, trans. Stuart Elden and Gerald Moore. London: Continuum.
Lenhart, Amanda, John Horrigan, Lee Rainie, Katherine Allen, Angie Boyce, Mary Madden, and Erin O'Grady. 2003. *The Ever-Shifting Internet Population: A New Look at Internet Access and the Digital Divide*. Pew Internet & American Life Project, April 16. www.pewinternet.org/2003/04/16/the-ever-shifting-internet-population-a-new-look-at-internet-access-and-the-digital-divide/
Lerman, Nina E, Ruth Oldenziel, and Arwen P. Mohun, eds. 2003. *Gender & Technology: A Reader*. Baltimore/London: The John Hopkins University Press.
Lewis, Michael. 1999. *The New, New Thing: A Silicon Valley Story*. New York: W. W. Norton & Co.
Licoppe, Christian. 2004. Connected Presence. *Environment and Planning D: Society and Space* 22, no. 1: 135–156.
Light, Ben, and Elija Cassidy. 2014. Strategies or the Suspension and Prevention of Disconnection: Rendering Disconnection as Socioeconomic Lubricant with Facebook. *New Media & Society* 16, no. 7: 1169–1184.
Lim, Sun Sun, ed. 2016. *Mobile Communication and the Family—Asian Experiences in Technology Domestication*. Dordrecht, Germany: Springer.
Ling, Rich, and Birgitte Yttri. 1999. *Nobody Sits at Home and Waits for the Telephone to Ring: Micro and Hyper-coordination Through the Use of the Mobile Telephone*. Perpetual Contact Workshop. New Brunswick, NJ: Rutgers University.

REFERENCES

Livingstone, Sonia. 2002. *Young People and New Media: Childhood and the Changing Media Environment*. London: Sage.

Livingstone, Sonia. 2003. Children's Use of the Internet: Reflections on the Emerging Research Agenda. *New Media and Society* 5, no. 2: 147–166.

Livingstone, Sonia. 2009. *Children and the Internet*. Cambridge: Polity Press.

Livingstone, Sonia, and Ellen J. Helsper. 2008. Parental Mediation of Children's Internet Use. *Journal of Broadcasting & Electronic Media* 52, no. 4: 581–599.

Mackay, Hugh, and Darren Ivey. 2004. *Modern Media in the Home: An Ethnographic Study*. Rome: John Libbey.

Mackenzie, Adrian. 2008. The Affect of Efficiency: Personal Productivity Equipment Encounters the Multiple. *Ephemera* 8, no. 2: 137–156.

Mackenzie, Adrian. 2010. *Wirelessness: Radical Empiricism in Network Cultures*. Cambridge, MA: MIT Press.

Mahar, Andrew. 2008. *Digital Inclusion Initiative*. Melbourne: Infoxchange. http://library.bsl.org.au/jspui/bitstream/1/2632/1/Assessing%20the%20economic%20benefits%20of%20digital%20inclusion.pdf

Manjoo, Farhad. 2009. Why 2024 Will Be Like Nineteen Eighty-Four. *Slate*, July 20. www.slate.com/articles/technology/technology/2009/07/why_2024_will_be_like_nineteen_eightyfour.html.

Mannell, Kate. 2017. De Certeau and Technology Resistance: Deceptive Texting as a Tactic of Everyday Life. *Platform* 8, no. 1: 40–55.

Marsh, Jackie, Lydia Plowman, Dylan Yamada-Rice, Julia Bishop, Jamal Lahmar, and Fiona Scott. 2018. Play and Creativity in Young Children's Use of Apps. *British Journal of Educational Technology* 49, no. 5: 870–882.

Marvin, Carolyn. 1988. *When Old Technologies Were New*. Oxford: Oxford University Press.

May, Jon, and Nigel Thrift, eds. 2001. *Timespace: Geographies of Temporality*. London: Routledge.

McCarthy, Anna. 2001. *Ambient Television: Visual Culture and Public Space*. Durham, NC: Duke University Press.

McCracken, Grant. 1988. *Culture and Consumption*. Bloomington: Indiana University Press.

McLean, Martin, and Natalia Voskresenskaya. 1992. Education Revolution from Above: Thatcher's Britain and Gorbachev's Soviet Union. *Comparative Education Review* 36, no. 1: 71–90.

Meyrowitz, Joshua. 1993. Images of Media: Hidden Ferment—and Harmony—in the Field. *Journal of Communication* 43, no. 3: 55–66.

Miller, Daniel. 2001. Possessions. In *Home Possessions: Material Culture and the Home*, edited by Daniel Miller, 107–122. Oxford: Berg.

Miller, Daniel. 2009. *The Comfort of Things*. Cambridge: Polity.

Miller, Daniel. 2012. *Consumption and its Consequences*. Cambridge: Polity.

REFERENCES

Miller, Daniel, and Don Slater. 2000. *The Internet: An Ethnographic Approach.* London: Berg.

Milne, Esther. 2012. *Letters, Postcards, Email: Technologies of Presence.* New York: Routledge.

Moores, Sean. 2005. *Media/Theory: Thinking About Media & Communications.* London: Routledge.

Moreland City Council. 2018. *Hard Waste Collection.* Melbourne: Moreland City Council. https://www.moreland.vic.gov.au/environment-bins/garbage-recycling-and-green-waste/hard-waste-collection/

Morley, David. 1980. *The "Nationwide" Audience: Structure and Decoding.* London: British Film Institute.

Morley, David. 1986. *Family Television: Cultural Power and Domestic Leisure.* London: Comedia Publishing Group.

Morley, D. 2003. "What's Home Got to Do with It?" Contradictory Dynamics in the Domestication of Technology and the Dislocation of Domesticity. *European Journal of Cultural Studies* 6, no. 4: 435–458.

Morley, David. 2007. *Media, Modernity and Technology: The Geography of the New.* London: Routledge.

Morley, David. 2009. For a Materialist, Non-Media-centric Media Studies. *Television & New Media* 10, no. 1: 114–116.

Morrison, Stacey L., and Ricardo Gomez. 2014. Pushback: Expressions of Resistance to the "Evertime" of Constant Online Connectivity. *First Monday* 19, no. 8. https://firstmonday.org/ojs/index.php/fm/article/view/4902

Murdock, Graham, Paul Hartmann, and Peggy Gray. 1992. Contextualising Home Computing: Resources and Practices. In *Consuming Technologies: Media and Information in Domestic Spaces,* edited by Roger Silverstone and Eric Hirsch, 146–160. London: Routledge.

Murray, Fergus. 2003. A Separate Reality: Science, Technology and Masculinity. In *Gendered Design? Information Technology and Office Systems,* edited by Eileen Green, 64–80. London: Taylor & Francis.

MyNetFone. 2016. 3.5 Million Teleworkers in Australia. *MyNetFone,* September 29. https://business.mynetfone.com.au/blog/3.5-million-teleworkers-in-australia

Nagy, Peter, and Gina Neff. 2015. Imagined Affordance: Reconstructing a Keyword for Communication Theory. *Social Media + Society* 1, no. 2. https://doi.org/10.1177/2056305115603385

Nancy, Jean-Luc. 1991. *The Inoperative Community,* trans. Peter Connor, Lisa Garbus, Michael Holland and Simona Sawhney. Minneapolis: University of Minnesota Press.

Nansen, Bjorn, Michael Arnold, Martin Gibbs, and Hilary Davis. 2009. Domestic Orchestration: Rhythms in the Mediated Home. *Time & Society* 18, no. 2: 181–207.

REFERENCES

Nansen, Bjorn, Michael Arnold, Martin Gibbs, and Hilary Davis. 2010. Time, Space and Technology in the Working-Home: An Unsettled Nexus. *New Technology, Work and Employment* 25, no. 2: 136–153.

Nansen, Bjorn, Michael Arnold, Martin Gibbs, and Hilary Davis. 2011. Dwelling with Media Stuff: Latencies of Materiality in Four Australian Homes. *Environment and Planning D* 29, no. 4: 693–715.

Nansen, Bjorn, Michael Arnold, Rowan Wilken, and Martin Gibbs. 2013a. *Broadbanding Brunswick: High-Speed Broadband and Household Media Ecologies: A Report on Household Take-Up and Adoption of the National Broadband Network in a First Release Site*. Sydney: Australian Communications Consumer Action Network (ACCAN).

Nansen, Bjorn, Kabita Chakraborty, Lisa Gibbs, Colin MacDougall, and Frank Vetere. 2012. Children and Digital Wellbeing in Australia: Online Mediation, Conduct and Competence. *Journal of Children and Media* 6, no. 2: 237–254.

Nansen, Bjorn, and Darshana Jayemanne. 2016. Infants, Interfaces, and Intermediation: Digital Parenting in the Production of "iPad Baby" YouTube Videos. *Journal of Broadcasting and Electronic Media* 60, no. 4: 587–603.

Nansen, Bjorn, Jenny Kennedy, Michael Arnold, Martin Gibbs, and Rowan Wilken. 2016. Digital Ethnographic Techniques in Domestic Spaces: Notes on Methods and Ethics. *Visual Methodologies* 3, no. 2: 86–97.

Nansen, Bjorn, Rowan Wilken, Michael Arnold, and Martin Gibbs. 2013b. Digital Literacies and the National Broadband Network: Competency, Legibility, Context. *Media International Australia* 147, no. 1: 18–28.

Nardi, Bonnie. 2010. *My Life as a Night Elf Priest: An Anthropological Account of the World of Warcraft*. Ann Arbor: University of Michigan Press.

Nardi, Bonnie A., and Vicki L. O'Day. 1999. *Information Ecologies: Using Technology with Heart*. Cambridge, MA: MIT Press.

Natalier, Kristin. 2003. "I'm Not His Wife": Doing Gender and Doing Housework in the Absence of Women. *Journal of Sociology* 39, no. 3: 253–269.

NetRatings Australia. 2005. *kidsonline@home: Internet Use in Australian Homes*. Sydney: Australian Broadcasting Authority.

Nevski, Elyna, and Andra Siibak. 2016a. The Role of Parents and Parental Mediation on 0–3-Year-Olds' Digital Play with Smart Devices: Estonian Parents' Attitudes and Practices. *Early Years: An International Research Journal* 36, no. 3: 227–241.

Nevski, Elyna, and Andra Siibak. 2016b. Mediation Practices of Parents and Older Siblings in Guiding Toddlers' Touchscreen Technology Use: An Ethnographic Case Study. *Media Education* 7, no. 2: 320–340.

Nikken, Peter. 2003. Parental Mediation of Children's Video Game Playing: A Similar Construct as Television Mediation. In *DiGRA '03—Proceedings of the 2003 DiGRA International Conference: Level Up*, Utrecht University, The Netherlands. http://www.digra.org/dl/db/05150.50493

REFERENCES

Nikken, Peter, and Jeroen Jansz. 2006. Parental Mediation of Children's Videogame Playing: A Comparison of the Reports by Parents and Children. *Learning, Media and Technology* 31, no. 2: 181–202.

Noble, David F. 1979. Social Choice in Machine Design: The Case of Automatically Controlled Machine Tools. In *Case Studies on the Labour Process*, edited by Andrew Zimbalist, 18–50. New York: Monthly Review Press.

Nye, David E. 1996. *American Technological Sublime*. Cambridge, MA: MIT Press.

OfCom. 2017. *Children and Parents: Media Use and Attitudes Report*. London/Oxford: OfCom/Oxford University Press.

Oudshoorn, Nelly, Els Rommes, and Marcelle Stienstra. 2004. Configuring the User as Everybody: Gender and Design Cultures in Information and Communication Technologies. *Science, Technology and Human Values* 29, no. 1: 30–63.

Packard, Vance. 1960. *The Waste Makers*. Philadelphia: David McKay Publishers.

Papert, Seymour. 1980. *Mindstorms: Children, Computers and Powerful Ideas*. New York: Basic Books.

Parikka, Jussi. 2011a. Media Ecologies and Imaginary Media: Transversal Expansions, Contractions, and Foldings. *Fibreculture Journal* 17: 34–50. seventeen.fibreculturejournal.org/fcj-116-media-ecologies-and-imaginary-media-transversal-expansions-contractions-and-foldings/

Parikka, Jussi, ed. 2011b. *Medianatures: The Materiality of Information Technology and Electronic Waste*. London: Open Humanities Press.

Parikka, Jussi. 2012. Zombie Media in *Leonardo*. Machinology Blog, September 5. https://jussiparikka.net/2012/09/05/zombie-media-in-leonardo/.

Parikka, Jussi, interviewed by Michael Dieter. 2014. New Materialism and Non-Humanisation. In *Speculative Realities (Blowup Reader # 6)*, edited by Michelle Kasprzak, 3–36. Rotterdam: V2 Institute for the Unstable Media.

Park, Sora, Catherine Middleton, and Matthew Allen. 2013. Conceptualizing the (Non) Users of the Internet. In *Proceedings of the Association of Internet Researchers 14th Annual Conference (IR14)*, October 23–26, Denver, CO.

Pasquier, Dominique. 2001. Media at Home: Domestic Interactions and Regulation. In *Children and Their Changing Media Environment: A European Comparative Study*, edited by Sonia Livingstone and Moira Bovill, 61–177. Mahwah, NJ: Lawrence Erlbaum.

Pickett, Steward T. A., and Mary L. Cadenasso. 2002. The Ecosystem as a Multidimensional Concept: Meaning, Model, and Metaphor. *Ecosystems* 5, no. 1: 1–10.

Pilcher, Jane. 1999. *Women in Contemporary Britain*. London: Routledge.

Pink, Sarah. 2004. *Home Truths: Gender, Domestic Objects and Everyday Life*. Oxford: Berg.

Plowman, Lydia, Joanna McPake, and Christine Stephen. 2010. The Technologisation of Childhood? Young Children and Technology in the Home. *Children & Society* 24, no. 1: 63–74.

REFERENCES

Portwood-Stacer, Laura. 2013. Media Refusal and Conspicuous Non-Consumption: The Performative and Political Dimensions of Facebook Abstention. *New Media & Society* 15, no 7: 1041–1057.

Postman, Neil. 2000. The Humanism of Media Ecology. In *Proceedings of the Media Ecology Association* 1: 10–16. http://www.media-ecology.org/publications/MEA_proceedings/v1/postman01.pdf

Prensky, Marc. 2001. Digital Natives, Digital Immigrants. *On the Horizon* 9, no. 5: 1–6.

Putnam, Robert D. 2000. *Bowling Alone: The Collapse and Revival of America Community*. New York: Simon & Schuster.

Rainie, Lee, Aaron Smith, and Maeve Duggan. 2013. Coming and Going on Facebook. *Pew Internet & American Life Project*, January 5. www.pewinternet.org/2013/02/05/coming-and-going-on-facebook/

Rakow, Lana F., ed. 1992. *Women Making Meaning: New Feminist Directions in Communication*. London: Routledge.

Rasmussen, Eric. 2017. Screen Time and Kids: Insights from a New Report. *PBS Kids for Parents*, October 19. http://www.pbs.org/parents/expert-tips-advice/2017/10/screen-time-kids-insights-new-report/

Reno, Joshua. 2015. Waste and Waste Management. *Annual Review of Anthropology* 44: 557–572.

Rheingold, Howard. 1995. Virtual Community and Civic Life in Amsterdam. *San Francisco Examiner*, September 5. https://people.well.com/user/hlr/tomorrow/amsterdam.html.

Rheingold, Howard. 2000. *The Virtual Community: Homesteading on the Electronic Frontier*. London: MIT Press.

Ribak, Rivka, and Michele Rosenthal. 2015. Smartphone Resistance as Media Ambivalence. *First Monday* 20, no. 11. https://firstmonday.org/ojs/index.php/fm/article/view/6307

Rogers, Everett M. 2003. *Diffusion of Innovations*, 5th ed. New York: Free Press.

Roose, Kevin. 2014. Why Are We Still Calling the Things in Our Pockets "Cell Phones"? *New York Magazine*, June 24. www.nymag.com/daily/intelligencer/2014/06/why-are-we-still-calling-them-cell-phones.html

Rosenblatt, Paul C., and Michael R. Cunningham. 1976. Television Watching and Family Tensions. *Journal of Marriage and the Family* 38, no. 1: 105–111.

Ruthven, Phil. 2012. *A Snapshot of Australia's Digital Future to 2050*. Melbourne: IBISWorld.

Rybczynski, Witold. 1986. *Home: A Short History of an Idea*. New York: Viking.

Sandvig, Cristian. 2013. The internet as infrastructure. In *The Oxford Handbook of Internet Studies*, edited by Willam H. Dutton, 86–108. Oxford: Oxford University Press.

Satchell, Christine, and Paul Dourish. 2009. Beyond the User: Use and Non-use in HCI. In *Proceedings of the 21st Annual Conference of the Australian*

Computer-Human Interaction Special Interest Group: Design: Open 24/7, November 23–27, University of Melbourne.

Saunders, Peter, and Peter Williams. 1988. The Constitution of the Home: Towards a Research Agenda. *Housing Studies* 3, no. 2: 81–93.

Schoenebeck, Sarita Yardi. 2014. Giving up Twitter for Lent: How and Why We Take Breaks from Social Media. In *Proceedings of the SIGCHI Conference on Human Factors in Computing Systems*, April 26–May 1, Toronto.

Scolari, Carlos A. 2012. Media Ecology: Exploring the Metaphor to Expand the Theory. *Communication Theory* 22, no. 2: 204–225.

Selwyn, Neil. 2006. Digital Division or Digital Decision? A Study of Non-users and Low-users of Computers. *Poetics* 34, no. 4–5: 273–292.

Serres, Michel. 1982. *The Parasite*, trans. Lawrence R. Schehr. Baltimore: Johns Hopkins University Press.

Shanteau, James. 1992. The Psychology of Experts: An Alternative View. In *Expertise and Decision Support*, edited by George Wright and Fergus Bolger, 11–23. New York: Plenum.

Shepherd, Chris, Michael Arnold, Craig Bellamy, and Martin Gibbs. 2007. The Material Ecologies of Domestic ICTs. *Electronic Journal of Communication* 17, no. 1–2. http://www.cios.org/www/ejc/v17n12.htm

Shepherd, Chris, Michael Arnold, and Martin Gibbs. 2006. Parenting in the Connected Home. *Journal of Family Studies* 12, no. 2: 203–222.

Shove, Elizabeth. 2003. *Comfort, Cleanliness and Convenience: The Social Organization of Normality*. New York: Berg.

Shove, Elizabeth. 2008. *Rushing Around: Coordination, Mobility and Inequality*. Lancaster: Lancaster University.

Shove, Elizabeth, Matthew Watson, Martin Hand, and Jack Ingram. 2007. *The Design of Everyday Life*. Oxford: Berg.

Silver, David. 2000. Margins in the Wires: Looking for Race, Gender, and Sexuality in the Blacksburg Electronic Village. In *Race in Cyberspace*, edited by Beth E. Kolko, Lisa Nakamura, and Gilbert B. Rodman, 133–150. New York: Routledge.

Silverstone, Roger. 1993. Time, Information and Communication Technology and the Household. *Time and Society* 2, no. 3: 283–311.

Silverstone, Roger, and Leslie Haddon. 1996. Design and the Domestication of Information and Communication Technologies: Technical Change and Everyday Life. In *Communication by Design: The Politics of Information and Communication Technologies*, edited by Roger Silverstone and Robin Mansell, 44–74. Oxford: Oxford University Press.

Silverstone, Roger, and Eric Hirsch, ed. 1992. *Consuming Technologies: Media and Information in Domestic Spaces*. London: Routledge.

Silverstone, Roger, Eric Hirsch, and David Morley. 1992. Information and Communication Technologies and the Moral Economy of the Household. In

Consuming Technologies: Media and Information in Domestic Spaces, edited by Roger Silverstone and Eric Hirsch, 15–31. London: Routledge.

Simmel, Georg. 2004. *The Philosophy of Money*, 3rd enlarged ed., ed. David Frisby, trans. Tom Bottomore and David Frisby. London: Routledge.

Slade, Giles. 2007. *Made to Break*. Cambridge, MA: Harvard University Press.

Sleeper, Manya, Alessandro Acquisti, Lorrie Faith Cranor, Patrick Gage Kelley, Sean A. Munson, and Norman Sadeh. 2015. I Would Like To . . ., I Shouldn't . . ., I Wish I . . . : Exploring Behavior-Change Goals for Social Networking Sites. In *Proceedings of the 18th ACM Conference on Computer Supported Cooperative Work & Social Computing*, March 15–18, Vancouver.

Solon, Olivia. 2017. "It's Digital Colonialism": How Facebook's Free Internet Service Has Failed its Users. *The Guardian*, July 27. https://www.theguardian.com/technology/2017/jul/27/facebook-free-basics-developing-markets

Spender, Dale. 1995. *Nattering on the Net: Women, Power and Cyberspace*. Melbourne: Spinifex.

Spigel, Lynn. 1992. *Make Room for TV: Television and the Family Ideal in Postwar America*. Chicago: University of Chicago Press.

Spigel, Lynn. 2001. Media Homes: Then and Now. *International Journal of Cultural Studies* 4, no. 4: 285–411.

Stillman, Larry, Michael Arnold, Martin Gibbs, and Chris Shepherd. 2010. ICT, Rural Dilution and the New Rurality: A Case Study of "WheatCliffs." *Journal of Community Informatics* 6, no. 2: 1–13.

Stonehenge. 1999a. DISR Report—Home System—Requirements Spec Ver 1-00.txt. Unpublished internal document, circulated June.

Stonehenge. 1999b. Home Brain Treatment v 01.txt. Unpublished internal document, circulated March 24.

Stonehenge. 1999c. Technology Diffusion Grant Feasibility Study PROJECT: Williams Bay Housing Development Cybersite. REPORT: Service Organisation Study and Implementation Plan. Unpublished internal document, circulated September.

Stonehenge. 1999d. Williams Bay Development: An Australian Pilot Research Project. Unpublished internal document, circulated September.

Strasburger, Victor C., and Edward Donnerstein. 1999. Children, Adolescents, and the Media: Issues and Solutions. *Pediatrics* 103, no. 1: 129–139.

Strate, Lance. 2004. A Media Ecology Review. *Communication Research Trends* 23, no. 2: 1–48.

Strate, Lance. 2006. *Echoes and Reflections: On Media Ecology as a Field of Study*. Cresskill, NJ: Hampton Press.

Strate, Lance. 2017. *Media Ecology: An Approach to Understanding the Human Condition*. New York: Peter Lang.

Symes, Colin. 1999. Chronicles of Labour: A Discourse Analysis of Diaries. *Time & Society* 8, no. 2: 357–380.

REFERENCES

Tacchi, Jo. 2006. Studying Communicative Ecologies: An Ethnographic Approach to Information and Communication Technologies (ICTs). In *Proceedings of the 56th Annual Conference of the International Communication Association*, Dresden, Germany. eprints.qut.edu.au/4400/1/4400_1.pdf

Thompson, Cadie. 2012. Facebook: About 83 Million Accounts Are Fake. *USA Today*, March 8. https://usatoday30.usatoday.com/tech/news/story/2012-08-03/cnbc-facebook-fake-accounts/56759964/1

Thompson, E. P. 1967. Time, Work-Discipline, and Industrial Capitalism. *Past & Present* 38, no. 1: 56–97.

Thrift, Nigel. 2007. *Non-representational Theory: Space, Politics, Affect*. London: Routledge.

Tichi, Cecelia. 1992. *Electronic Hearth: Creating an American Television Culture*. Oxford: Oxford University Press.

Toffler, Alvin. 1980. *The Third Wave*. Toronto: Bantam Books.

Tolmie, Peter, Andy Crabtree, Tom Rodden, Chris Greenhalgh, and Steve Benford. 2007. Making the Home Network at Home: Digital Housekeeping. In *ECSCW 2007: Proceedings of the 10th European Conference on Computer-Supported Cooperative Work*, Limerick, Ireland, September 24–28, edited by Liam J. Bannon, Ina Wagner, Carl Gutwin, Richard H. R. Harper, and Kjeld Schmidt, 331–350. London: Springer-Verlag.

Trulove, James G. 2002. *The Smart House*. New York: HDI.

Tufekci, Zeynep. 2008. Grooming, Gossip, Facebook and MySpace: What Can We Learn About These Sites from Those Who Won't Assimilate? *Information, Communication & Society* 11, no. 4: 544–564.

Turkle, Sherry. 2017. *Alone Together: Why We Expect More from Technology and Less from Each Other*. London: Hachette.

Valcke, Martin, Sarah Bonte, Bram De Wever, and Isabel Rots. 2010. Internet Parenting Styles and the Impact on Internet Use of Primary School Children. *Computers & Education* 55, no. 2: 454–464.

van Dijck, Jose. 2013. *The Culture of Connectivity: A Critical History of Social Media*. New York: Oxford University Press.

Venkatesh, Alladi. 1996. Computers and Other Interactive Technologies for the Home. *Communications of the ACM* 39, no. 12: 47–54.

Virilio, Paul. 2006. *Speed and Politics*, new ed., trans. Mark Polizzotti. Cambridge, MA: MIT Press.

Wajcman, Judy. 1994.Technological A/Genders: Technology, Culture and Class. In *Framing Technology: Society, Choice and Change*, edited by Lelia Green and Roger Guinery, 3–14. St. Leonards, NSW: Allen & Unwin.

Wajcman, Judy. 2004. *TechnoFeminism*. Cambridge: Polity.

Wajcman, Judy, Michael Bittman, Paul Jones, Lynne Johnstone, and Jude Brown. 2007. *The Impact of the Mobile Phone on Work/Life Balance*. Canberra: Australian National University.

REFERENCES

Wajcman, Judy, and Nigel Dodd, eds. 2016. *The Sociology of Speed: Digital, Organizational and Social Temporalities*. Oxford: Oxford University Press.

Wakkary, Ron, and Leah Maestri. 2008. Aspects of Everyday Design: Resourcefulness, Adaptation, and Emergence. *International Journal of Human–Computer Interaction* 24, no. 5: 478–491.

Ward, Katie. 2005. Internet Consumption in Ireland—Towards a "Connected" Life. In *Media, Technology and Everyday Life in Europe*, edited by Roger Silverstone, 107–123. Aldershot: Ashgate.

Warschauer, Mark. 2003. *Technology and Social Inclusion: Rethinking the Digital Divide*. Cambridge, MA: MIT Press.

West, Darrell M. 2010. *An International Look at High-Speed Broadband*. Washington, DC: Brookings Institution. http://www.brookings.edu/~/media/research/files/reports/2010/2/23%20broadband%20west/0223_broadband_west

Wheelock, Jane. 1992. Personal Computers, Gender and an Institutional Model of the Household. In *Consuming Technologies: Media and Information in Domestic Spaces*, edited by Roger Silverstone and Eric Hirsch, 97–112. London: Routledge.

Wilken, Rowan, Michael Arnold, and Bjorn Nansen. 2011. Broadband in the Home Pilot Study: Suburban Hobart. *Telecommunications Journal of Australia* 61, no. 1: 5.1–5.16.

Wilken, Rowan, Bjorn Nansen, Michael Arnold, Jenny Kennedy, and Martin Gibbs. 2014. National, Local and Household Media Ecologies: The Case of Australia's National Broadband Network. *Communication, Politics & Culture* 46, no. 2: 136–154.

Williams, Raymond. 1975. *Television: Technology and Cultural Form*. London: Fontana.

Winner, Langdon. 1986. *The Whale and the Reactor: A Search for Limits in an Age of High Technology*. Chicago: University of Chicago Press.

Wolf, Naomi. 1990. *The Beauty Myth*. London: Chatto & Windus.

Woodstock, Louise. 2014. Media Resistance: Opportunities for Practice Theory and New Media Research. *International Journal of Communication* 8: 1983–2001.

Wright, Philippa. 2005. *A Community Intranet: Factors Affecting the Establishment of Information Communications Technologies at the Neighbourhood Level*. PhD thesis, University of Melbourne.

Wyatt, Sally. 2003. Non-Users Also Matter: The Construction of Users and Non-Users of the Internet. In *How Users Matter: The Co-construction of Users and Technologies*, edited by Nelly Oudshoorn and Trevor Pinch, 67–79. Cambridge, MA: MIT Press.

Zerubavel, Eviatar. 1985. *The Seven Day Circle: The History and Meaning of the Week*. New York: Free Press.

Zuboff, Shoshana. 2019. *The Age of Surveillance Capitalism: The Fight for a Human Future at the New Frontier of Power*. London: Profile Books.

INDEX

Figures are indicated by *f* following the page number

For the benefit of digital users, indexed terms that span two pages (e.g., 52–53) may, on occasion, appear on only one of those pages.

accommodation of technology, 86–87, 90–91, 93–96, 99–100, 101
    mutual, 143–44
ADSL technology, 45
Altheide, David, 8–9, 16, 59, 68–69
always on technology, 185–86, 193–94
Ananny, Mike, 204–5
answering machines, 241–43, 242*f*
Apple TV, 50–51, 244
appropriation of technology, 9–10, 86–126
apps, 208–9
articulation, 10, 274–75
Australian Centre for Broadband Innovation (ACBI)
    Broadband Connected Homes Study, 41–43
Australian Communication and Media Authority (ACMA), 44–45, 51–52

Banks, David, 206
Barlow, John Perry, 28–29
basic phone, 208–9
Bausinger, Hermann, 63–64

bedroom
    culture, 18, 48–49, 61–62
    mobile devices in, 32–33, 108–12, 180–81
    as no-go zone, 108
    See also no-go zones
Blacksburg (US), 3–4, 25
Blunt, Alison, 66
Brand, Stewart, 20–21, 28–29
broadband, 9–10, 86–88
    appropriation by household, 116–26
    architecture, 49–50, 114
    link between household composition and bandwidth, 121, 213–14
    need to upgrade, 52
    visions of infrastructure, 39–45
    See also National Broadband Network (NBN)
broadcasting, 16–17
    See also radio, television
Brown, Steven, 232
brown goods, 56–57, 137
    See also gendering of technology; white goods

301

INDEX

Brubaker, Jed, 204–5
Brunswick sample, 6
bullying. *See* cyberbullying
Burns, Alex, 43

cable networks, 9–10
Cadenasso, Mary, 67
Callon, Michel, 223
Case, Steve, 37–38
Cat 5 cabling, 26, 230
CDs, 19–20
Chicago School, 67
cloud computing, 9–10, 51–52, 87–88
Cockburn, Cynthia, 11, 128, 139
communitarianism, 29–30
community
    changing conceptions, 28–29
    impact of media ecologies, 28–30
    See also community digital networks
community digital networks, 29–31
    community intranet, 31, 222–24, 270
    creation of social ties, 30
computer. *See* home computing; mobile computing
Connected Homes Project, 4, 5–6, 107, 116, 180–81, 211, 218, 244–45
    Annie, Robert and Andy, 164–70, 171, 178–79
    Annika, 186
    Eileen, 187–88
    Georgie, 120, 188
    Jacka, 108, 211
    June, 104
    Katie and Bob, 115, 119, 198, 248–49
    Mary and John, 37–39, 97, 98–99, 103, 130–43, 186, 198, 218–19, 228
    Maurice, 185–86
    Sam, 188
    Sharon, 115
    Tess and James, 195
    Tom, 102–3, 119–20
    use of domestic probe, 75–79
    Yukiko, 188
    See also domestic probe
connectivity, 20
    See also networked home
consumerism, 90–91, 239
consumption
    mass, 91–92
    sociology of, 60
Crabtree, Andy, 74
Crawford, Kate, 204–5
cultural geography, 65
cultural probe, 74
    See also domestic probe
cultural proprioception, 85
curbside hard rubbish collection, 241–43, 242f, 252–53
cyberbullying, 33, 167–68, 275–76

de Certeau, Michel, 205
desktop computer. *See* home computing
dial-up internet, 19–20, 273–74
diffusion of technology, 63, 86–87, 90–92
    early and late adopters, 91–92
    inevitability, 91–92
    static ontology, 91–92
digital cameras, 9–10
digital divide, 43–44
digital domesticity, 71–72
    approach to research, 1–2
    defined, 1
    entertainment aspects, 56–57
    eras, 4–6, 273–74
    longitudinal change, 3
    streaming, 9–10, 87–88
    See also Connected Homes Project; High-Speed Broadband Project; Wired Homes Project
digital economy, 116–17
digital expertise, 129–30
    characteristics, 129–30
digital home. *See* networked home
digital housekeeping, 10, 127–63, 228–29, 275
    contribution to self-worth, 158
    distribution of expertise, 155–60
    expertise and, 128–30, 134, 136–43, 146–47, 151–55
    gender relations, 127, 128, 130–36, 137–38, 144, 155
    identity and, 11
    identity and, 136–43
    labor and conditions of expertise, 161–63
    managing devices, 149–50
    managing digital content, 145–47

302

INDEX

managing digital networks, 148–50
nature of expertise, 156
origin of term, 128–29
power relations, 127
practices, 143–45
substitution for other domestic tasks, 157–58
digital natives, 11
vs digital immigrants, 128, 129–30
digital radio, 9–10
See also radio
Digitale Stad initiative (Amsterdam), 3–4, 25
dining room, 111
displacement of technologies, 14–15, 237–68, 277–78
discarding devices, 251–56
dispositifs, 90–91
Dodd, Nigel, 184–85
domestic co-habitation, 66
domestic media ecology, 1–2, 16, 50, 86, 271
dense, 90–91
major changes, 271
technology rich, 44–45
See also domestic media technologies; networked home
domestic media ensembles, 63–64
domestic media technologies
accumulation, 237
aggregation, 95–96
co-location, 100–2
coordination, 196–99
domestic rhythms, 184–200
family politics, 180
fluid evolution, 25
gendering, 11
histories, 7–8, 11–12, 16–55
interaction, 269–70
interpersonal relations and, 1–2, 86
life cycles, 1–2
material agency, 101
mobility, 65–78
monitoring of use, 108–10, 143–44
motivations for, 10
multiple devices, 44–45, 98–101, 120, 177–78, 238
rapid turnover, 90–91

role in parent–child interaction, 174–76
sentimental value, 14–15, 237, 262–63
sharing, 15
techno-dystopian critique of, 216–17
temporal impact, 184–200
See also digital housekeeping; networked home; smart home; technological shift; temporal aspects of domestic media technologies
domestic probe, 5–6, 59–60, 73–74, 81–83, 165, 273–74
Connected Homes Project, 75–79
diary room entry, 79–81
document-box, 76
EthnoCorder, 79, 79*f*, 81
frustro-meter, 76
High Speed Broadband Project, 75–79, 82–83
householders as co-researchers, 81–82
inspiration trap, 76
iPad minis, 79
junk mail catalogue, 76–78
Missions Impossible, 77
nature documentary, 79
news report, 79
objects to think with, 81–82
paparazzi shot, 79–81, 80*f*
random sampler, 76
research schedule, 82–83
researchers in absentia, 81–82
special purpose camera, 76–77
talk show interview, 79–81
technology tour, 82–83, 273–74
televisual tasks, 79
domestication approach, 59–64
appropriation, 60–61
conversion, 60–61
to digital home, 63–64
incorporation, 60–61
objectification, 60–61
studies of media homes, 61–62, 94–95
domesticity, 1, 99–100
beyond domestication, 64–66
de-domestication, 63–64
dislocation, 61–62
orchestrating, 195–200
re-domestication, 63–64
See also home

INDEX

Downes, Toni, 173
Dropbox, 51–52, 229
Duffy, Frank, 21–22
DVDs, 101, 112–14, 130–31, 133–34, 175–76, 177, 180–81, 249

e-commerce, 9–10
e-Key, 22, 24, 25, 26
ecological approach, 1–2, 8–9, 67, 88, 273
ecosystems of technology, 70
email, 51–52
    as storage space, 232
embodied media, 54
Etzioni, Amitai, 28–29

Facebook, 230–31
    fake accounts, 230–31
    privacy concerns, 221
    real name policy, 230–31
    vacations, 212
family
    evolution of, 20–21
    synchronization, 20
fax machine, 232–34, 241–43
feature phone, 208–9
Flaherty, Michael, 193–94
flat-screen TVs, 35, 90–91
Flickr, 51–52
Foucault, 90–91
Free Basics movement, 202–3
free-to-air television, 19–20
Friendster, 230–31
Fuller, Matthew, 59, 68, 69
Fürst-Dilić, Ruža, 139–41

gaming, 33–34
    concerns about, 33–34
    consoles, 9–10, 90–91
gated communities, 28
Gaver, Bill, 74
gendered socialization, 137–38
gendering of technology, 58, 278–79
gesture-controlled interfaces, 9–10, 87–88
Gilmore, John, 28–29
global financial crisis
    infrastructure projects, 116
Gomez, Ricardo, 212

government service provision online. *See* online government service provision
gramophone, 16
Gregson, Nicky, 239

habitation, 65
    co-habitation, 65, 66
    See also home
Hand, Martin, 88–90
Haring, Kristen, 137–38
Hertz, Garnet, 249
Hetherington, Kevin, 263–64
High-Speed Broadband Project, 6, 241–43
    Adele and Tom, 157, 227–28
    Andrew, 208–9
    Andy, 123
    Antonio, 151–52
    Beverley, 246–47, 257–59
    Carl, 148–49, 246–47
    Christine, 146–47, 151–52
    co-located devices, 100
    Craig, 263f
    Dawn, 111–12
    Deborah and Donald, 152–53, 220
    Dennis, 243
    Diane and Scott, 152, 153–55
    Doug, 145–46
    Eliza, 246–47
    Gloria, 146–47
    Graham, 101
    Grant, 119
    Howard, 52–53
    Ivan, 249
    Jaume, 214–15
    Jens and Jenny, 261–62
    Jeremy, 147, 158, 247
    Joel and Sam, 214
    Juliette, 225
    Malcolm, 46, 260
    Malcolm and Nysha, 99, 159–60, 246–47, 265
    Matthew, 265–67
    Michael and Anne, 123, 149–50, 264
    Natasha, 120
    Odette, 111
    parental strategies, 171–72
    Peter, 124

INDEX

Riley and Ashley, 46, 147, 151–52, 157, 158–59, 264
Sada, 117–18
Shawn and Christine, 260–61
Simon, 53
Stephanie and Peter, 224–25, 226
use of domestic probe, 79–81
See also domestic probe
Hirsch, Eric, 56
Hobart sample, 6, 43
home
  act of dwelling, 88, 89–90
  appliances, 90
  architecture, 49, 88
  assemblages of, 10
  command and control centre, 57–58
  consumption centre, 58
  data-processing centre, 57–58
  fluid and evolving institution, 70–71, 272–73
  immaterial labor, 162–63
  living with things, 65
  machine for living, 270
  materiality, 1–2, 20–21, 88, 263–64
  multifunctionality, 89–90
  nodal point in networked infrastructure, 20, 48–49
  non-human occupants, 88–89
  open-plan, 49
  place of leisure, 56–57
  private vs public space, 70–71
  services, 21–22
  setup of, 10
  site, 21–22
  site of production, 20, 57–58, 115
  site of reproduction, 20
  skin, 21–22
  sophisticated entertainment centre, 36–37, 56–57
  space-plan, 21–22
  stuff, 21–22, 88–89, 225–26
  See also digital housekeeping; domestic media technologies; living room; networked home
home banking, 58
Home Brain, 22–28, 31, 37, 270
  evolution of technology, 25–28

Home Information System, 22
Home Linking System, 23
Home Management System, 22
  virtual fridge door, 26
home computing, 16, 18, 19–20, 90–91, 93–94
  influence on household relations and imaginaries, 61–62
  positioning of computer, 180–81
home office, 9–10, 102–4, 105f, 273–74
  technology graveyard, 103, 104
home shopping, 58
home theatre, 18, 56–57
human–computer interaction (HCI), 204
human ecology, 67
humanism, 68

infomated household, 70
information superhighway, 19–20
internet, 16
  impact on family life, 32, 61–62
Internet Governance Forum, 203
intimate histories, 71–72, 273–74
iPad, 50, 210–11
iPhone, 50, 210–11. *See also* mobile phone; smartphone
ISDN (integrated services digital network) connection, 19–20

Japanese Multifunction Polis, 12–13

Kaczynski, Ted, 217–18
Kapor, Mitch, 28–29
Key4IT, 22, 25, 26–28, 27f
Kim, Sunyoung, 248, 256–57, 263–64
Kinect, 251–52
kitchen as contested space, 110–11
Kurzweil, Ray, 28–29

Lally, Elaine, 85
landline telephone, 16, 132–33
  disruption to family life, 32–33
  shift to mobile technology, 240–43
  shift from POTS to VoIP, 238, 241–43
laptop computer, 107–8. *See also* mobile devices
Latour, Bruno, 69–70, 243

INDEX

Leach, Tessa, 251–52
Lee, Heijin, 191
Liebenau, Jonathan, 191
Lightfoot, Geoffrey, 232
living room, 96–102
    changes due to television, 16–17, 18
    impact of mobile devices, 96–97
    impact of wireless connectivity, 96–97, 98–99
    new furniture, 16–17
    new ways of eating, 18
living together separately, 18, 61–62, 98, 110

Mannell, Kate, 205
Marvin, Carolyn, 16, 32–33
mass marketing, 91–92
materiality
    appropriation of artifacts, 65
    living with things, 65
    material anthropology, 65
    material culture studies, 65
    material turn in social sciences, 66
    media, 94–95
    quasi-object, 66
May, Jon, 195
McGrail, Stephen, 43
McLuhan, Marshall, 67
media and communication ecologies, 66–70
media for development, 69
media resistance, 208–9, 210
media room, 105–8
    disappearance of, 106–8
    home cinema, 105–6
media-saturated environment, 44–45
Melbourne sample, 6
Merlin, 22, 25, 26
Mesh Potato, 160
Mitchell, William, 28–29
mobile computing, 9–10, 87–88
mobile devices, 273–74
    breakages, 245–47
    degraded performance, 246–48
    disposal, 252–54
    impact on domestic space, 96–97, 107–8, 109–10
    individual nature, 179, 226–27
    multiple, 210–11, 225–26, 238
    recycling, 254
    See also mobile phones; domestic media technologies
mobile media, 48–55
    multiple screen use, 50–51
    See also domestic media technologies; mobile phone; tablets
mobile phone, 50, 90–91
    influence on family life, 61–62
    obsolescence, 245–46
    spare emergency phone, 261–62
    stockpiling, 256–57, 263–68
    See also iPhone; smartphone
moral panic, 32
Morley, David, 63–64
Morrison, Stacy, 212
MSN, 35, 164–66, 170
Murray, Fergus, 136–37
music formats, 219–20

Nagy, Peter, 208
National Broadband Network (NBN), 6, 39–40, 116–26
    compensation for poor, 213–14
    early adopters, 121
    fiber-to-the-node vs fibre-to-the-premises, 40–41, 118
    increasing importance, 42–43
    infrastructure, 121–24
natural user interface (NUI), 53
Neff, Gina, 208
negotiation of domestic media technologies, 12, 164–200
    domestic rhythms, 184–200
    See also parenting
net-nanny software, 32–33
Netville (Toronto), 12–13
networked home, 31–39, 45–48
    cost of maintaining, 129–30
    ecology of communication, 48–49
    flexible use, 110
    hidden infrastructure, 46–47, 112–14
    multiple devices, 44–45, 49–50, 177–78, 225–26
    node in global network, 109–10
    wireless router, 46–47, 90–91, 121–24

INDEX

See also domestic media technologies; smart home
newspapers, 19–20
no-go zones, 86–87, 108–12, 193, 211
non-human, 66
non-use of technologies, 13, 200, 201–36, 276
   active resistance, 207, 216–21
   contexts of, 206
   cyclical relations, 205–6, 212
   deficit model, 202–3, 204
   form of use practice, 202
   future users concept, 202–3
   intensity, 13–14
   latent users, 203
   partial use, 14, 108–9, 193–94, 207, 208–16, 276
   passive neglect, 207, 222–26
   peripheral, 14
   peripheral use, 227–28
   practices, 206–36
   privacy concerns, 214–15, 221
   pushback, 212
   radical, 14, 276
   radical use, 207, 230–36
   rejection, 217–18
   related to inequality and exclusion, 205
   resistance, 217–18, 221, 276
   resistance to music formats, 219–20
   technical contingency approach, 212–13
   use by proxy, 230, 276
   vicarious use, 14, 207, 226–30

obsolescence, 14–15, 126, 239, 240–45, 277–78
   breakages, 245–47
   environmental consequences, 15
   planned, 240
   precious junk, 138–39, 262–63
   psychological, 240, 243
   technological, 240
   unused devices, 225–26
one laptop per child movement, 202–3
online government service provision, 9–10, 87–88
   resistance to, 217–18

Ormrod, Susan, 11, 128

parenting, 12, 275–76
   concern about internet use, 164–70, 188
   evolution of approaches, 34–35
   impact of internet, 33
   managing time, 12
   negotiation of domestic media technologies, 12, 174–76
   parental controls, 33–34, 149–50, 178–79
   parental mediation, 172–73
   placement of technologies, 34–35
   strategies, 34, 171–84
   See also screen time
Parikka, Jussi, 69–70, 249
participation gap, 43–44
Paulos, Eric, 248, 256–57, 263–64
personal computer. *See* home computing
personal entertainment devices, 9–10
Pickett, Stuart, 67
post-human, 66
post-social, 66
Postman, Neil, 67–68
Prensky, Marc, 11, 128
private consumption, 28–29
   link to multiple devices, 49–50
   See also consumption
private sphere, 28–29
   retreat to, 48–49
privatization of entertainment, 16–17, 56–57
Putnam, Robert D., 28–29

radio, 16
   replacement by television, 16–17, 93–94
   See also digital radio
Rainie, Lee, 212
repurposing of technologies, 14–15, 245–50, 277–78
   dysfunction, 245–50
   passing on, 14–15, 237, 239, 254
   reselling, 254–56
   stockpiling, 256–68, 277–78
research challenges and innovations, 70–83
research method, 6–7, 273–74
   digital ethnographies, 72–74
   household media ethnographies, 72–73

INDEX

research method (*cont.*)
  literature survey, 6–7
  participant observation, 71–73
  single technologies focus, 72–73
  snapshot approach, 72–73
  staged process, 73–74
  technology tour, 5–6
  See also domestic probe; domestication approach to research
reverse salient, 243–44
Rheingold, Howard, 28–29
Ribak, Rivka, 208–9
Rosenthal, Michele, 208–9
Ruthven, Phil, 54, 118
Rybczynski, Witold, 88–89

Schoenebeck, Sarita Yardi, 212
screen time, 181–83
  parental concern about, 183–84
Seipp-Williams, Lucas, 193–94
selective computer use, 204
  See also non-use
semiotic approaches to domestic life, 66
Shove, Elizabeth, 88–90, 195–96, 198–99
Silverstone, Roger, 56
Simmel, Georg, 184–85
Siri, 185–86
Skype, 120–21
Slade, Giles, 240
sleep deprivation, 275–76
smart home, 1, 18, 26, 56–57, 58, 61–62, 87–88
  entertainment hub, 37
  future-proofing, 37
  highly networked, 42–43
  infrastructure, 112–15
  See also domestic media technologies; home; networked home; virtual assistants
smart televisions, 9–10, 87–88
  See also television
smartphone
  mobile media access, 39–40
  partial use of features, 209
  software bloat, 210
  See also iPhone

social isolation, 28–29
social media, 230–31
  See also Facebook
social networks, 9–10
socio-technical relations, 2
Southerton, Dale, 88–90
Special Rapporteur on the Promotion and Protection of the Right to Freedom of Opinion and Expression, 203
Spigel, Lynn, 1, 16–17, 88–89
streaming, 9–10, 87–88, 93–94, 273–74
Symes, Colin, 195–96

tablets, 50, 120
Tacchi, Jo, 68, 69
techno-utopianism, 29–30, 40
technological shift, 9–10
  from limited to multiple, 3
  from wired to mobile, 3
technological sublime, 35–36
technology-free zones. *See* no-go zones
telephone. *See* landline telephone; mobile phone; smartphone
television, 16
  analog to digital transition, 238, 240–41
  cathode-ray, 87–88, 238
  changes to home, 16–17, 18, 88–89, 93–94, 272
  collective viewing, 61–62
  colonization of public space, 63–64
  electronic hearth, 16–17
  flat-screen, 34–35, 98, 110
  migration of older TV sets, 98
  parental concerns about, 176–77
  role in family interaction, 175–76
  social medium, 16–17
  TV dinners, 93–94
  viewing location, 97
  viewing times, 94
  See also living room; smart televisions
temporal aspects of domestic media technologies, 12–13, 184–200
  asynchronous consonance, 12–13, 184–85, 191–95
  impact of digital technologies, 12–13
  orchestrating domesticity, 12–13, 184–85, 195–200

polychronic dissonance, 12–13, 184–85, 189–91
polyphonic drone, 12–13, 184–88
temporal mediation, 211–12
See also domestic media technologies; negotiation of domestic media technologies
Thrift, Nigel, 195
Toffler, Alvin, 37–38
Tolmie, Peter, 128–29, 155–56
touchscreen media, 53

UseNet Newsgroups, 19–20

Virilio, Paul, 184–85
virtual assistants, 54, 87–88, 112
voice recognition software, 87–88
VoIP, 50–51

Wacjman, Judy, 184–85, 192
Warschauer, Mark, 43–44
white goods, 58, 128, 137
See also brown goods; gendering of technology

Wi-Fi, 273–74
  impact on home technology use, 244–45
Williams Bay, Melbourne, 3–6, 24–28
Wired Homes Project, 4–5, 98–99, 180–81, 211, 218, 222–24
wireless internet
  impact on domestic space, 9–10, 18, 108
Wolfe, Naomi, 142
Woodstock, Louise, 208–9
work–life balance, 40–41
working from home, 37–38, 102–4
  telework, 38, 117
  work–home boundary, 48–49, 103
  See also work–life balance
World Summit on the Information Society, 203
Wyatt, Sally, 217–18

YouTube, 51–52, 98–99

Zerubavel, Eviatar, 193–94
zombie media, 15, 238, 249